Vibrational Spectroscopy: Theory and Applications

Vibrational Spectroscopy: Theory and Applications

Edited by **Hugo Kaye**

New York

Published by NY Research Press,
23 West, 55th Street, Suite 816,
New York, NY 10019, USA
www.nyresearchpress.com

Vibrational Spectroscopy: Theory and Applications
Edited by Hugo Kaye

International Standard Book Number: 978-1-63238-461-4 (Hardback)

Printed in the United States of America.

Contents

Preface

This book presents the theory as well as applications of vibrational spectroscopy with the help of state-of-the-art information. Diverse spheres including chemistry, physics, astronomy, medicine, mineralogy and biology have found use of infrared and Raman spectroscopy. This text provides some instances of the uses of vibrational spectroscopy in supramolecular chemistry, inorganic chemistry, solid state physics and other such fields, including those of molecule-based materials or organic-inorganic interfaces.

After months of intensive research and writing, this book is the end result of all who devoted their time and efforts in the initiation and progress of this book. It will surely be a source of reference in enhancing the required knowledge of the new developments in the area. During the course of developing this book, certain measures such as accuracy, authenticity and research focused analytical studies were given preference in order to produce a comprehensive book in the area of study.

This book would not have been possible without the efforts of the authors and the publisher. I extend my sincere thanks to them. Secondly, I express my gratitude to my family and well-wishers. And most importantly, I thank my students for constantly expressing their willingness and curiosity in enhancing their knowledge in the field, which encourages me to take up further research projects for the advancement of the area.

Editor

Vibrational Spectroscopy of Complex Synthetic and Industrial Products

Mario Alberto Gómez and Kee eun Lee
McGill University, Department of Materials Engineering
Canada

1. Introduction

Arsenic and its adverse effects on the environment and society impose a serious risk as it can be observed in cases such as India, Bangladesh, Vietnam, Nepal, and Cambodia. These problems can arise from geochemical processes but also man-made activities like mining and processing of minerals (Copper, Gold, Cobalt, Nickel and Uranium) that may further contribute to the risk of environmental disasters. In either case, problems and environmental disasters maybe prevented if a detailed chemical identification and understanding of the arsenic species and their chemical properties such as solubility is understood. The processing of these mineral ores (containing various metals) generates arsenic-containing solid wastes which are disposed in tailings management facilities outside in the environment (Riveros et al., 2001; Dymov et al., 2004; Defreyne et al., 2009; Mayhew et al., 2010; Bruce at al., 2011). Most often the phases that are produced after the process is completed is often a multicomponent in nature and contains a variety of elemental and individual phases (often $\geq$ 3-10 phases are expected to be present at the end of the process) produced produced along the chemical process. To complicate the understanding of these phases even more, depending upon the processing conditions, the produced unwanted materials (from an economic and recyclable aspect) may also be of a poorly crystalline nature (nano-scale order) in addition to being multi-component in nature and as such, simple tasks such as identification and then inferring understanding of its chemical properties (such as arsenic release into the environment) becomes very difficult if not nearly impossible. In general, as a result of their multicomponent nature, and perhaps due to historical relation, the use of lab based XRD has been the main tool to tackle the identification of these phases via the use of peak matching and Rietveld type of phase fitting analysis. Recently, synchrotron based soft and hard X-ray techniques (XANES and EXAFS) have been used in a similar fashion to overcome the poorly crystalline (lack of long range order) and multicomponent barrier of these industrially produced complex samples using similar types of mathematical types of fitting routines (Principal Component Analysis and Target Transformation) which still have limitations in identifying mixtures (>2-3) of similar phases with various coordination numbers/environments for the same probing atom of interest.

The need to develop alternative energy storage devices has become one of the main focuses in North America and around the World as a result of the fact that as the population of the world increases, our oil energy supplies are steadily and quickly decreasing with time.

Therefore intensive research, publications and funding have been mainly invested in Hydrogen energy, Li-ion types of batteries and Photovoltaic types of energy cells, although geothermal and wind type of alternatives are also being explored, the latter remain the focus for an alternative energy solution. To overcome technological plateaus in the development of technologies for energy (Peter, 2011; Scrosati & Garche, 2010), a deeper understanding of the processes, interactions and properties of the materials is needed to tailor the material to our benefit. However, in reality, the overall identification of the electrochemical process and physical aspects of the materials is nearly impossible due to their multicomponent and complex nano structure. As a result a number of multiple analysis techniques and methods such as X-ray and electron tomography, diffraction, and spectroscopy's, have been employed to investigate the structural and electronic nature of these types of systems. Yet, the complex nature of the multicomponent systems in devices such as Li-ion batteries or Dye Sensitized Solar Cell (DSSC) Photovoltaic's still push the boundaries of the understanding (much that still remains unknown or in some cases debated) of these systems and the analysis methods/techniques that are employed.

In this chapter, no detailed outline of the vibrational techniques, no instrumentation or theory is discussed in this chapter rather the reader is referenced to the numerous published works in journals and books on these subjects from various publishers (Harris and Bertolucci, 1989; Coates, 1998; Hollas, 2004; Nakamoto, 2009). Rather, this chapter describes three individual case studies in two different research areas: Arsenic species formed during the processing of minerals and the biding interactions in Dye Sensitized Solar Cells. In these sections, a description of the importance of the problem at hand and the use/application of vibrational spectroscopy in correlation with other information from other complementary techniques or studies is reported. From these case studies, the great advantages and most importantly disadvantages/limitations of the use of vibrational spectroscopy arise (in addition to other techniques). This is especially true in the case of muti-component systems such as those found in nature or industrially produced materials for which vibrational spectroscopy and other techniques (even the standard XRD) become limited.

2. The synthetic and natural Ca(II)-Fe(III)-AsO₄ (yukonite & arseniosiderite) system

2.1 Introduction

The hydrothermal $Ca(II)-Fe(III)-AsO_4$ system has been actively investigated (Swash, 1996; Becze & Demopoulos, 2007; Becze et al., 2010) given its relevance to precipitation that occurs in industrial arsenical processing liquors (such as those found in the Uranium industry in Northern Saskatchewan, Canada) upon neutralization with lime (CaO). More recently, the presence of these types of phases has gained interest as a result of their formation as secondary mineral in the mining tailing operations of Gold operations.

Among the various calcium ferric arsenate minerals out there, yukonite and arseniosiderite are the most common ones. Yukonite has been reported to occur most often as fractured, gel-like aggregates of dark brownish color. The chemical composition of this rare mineral has been reported to vary substantially in terms of its Fe, Ca, As and H₂O content, as can be observed from previous reports (Tyrrell & Graham 1913; Jambor, 1966; Dunn, 1982; Swash, 1996; Ross & Post 1997; Pieckza et al., 1998; Nishikawa et al., 2006; Paktunc et al., 2003, 2004; Becze & Demopoulos 2007; Walker et al., 2009; Garavelli et al., 2009). X-ray diffraction analysis of this material gives broad reflections typical of a mineral with poorly ordered (semi-crystalline) structure (Ross & Post 1997; Garavelli et al., 2009) while electron diffraction measurements (Nishikawa et al., 2006) suggested single crystal diffraction domains with an orthorhombic or hexagonal symmetry. As K-edge EXAFS measurements (Paktunc et al., 2003, 2004) proposed a local molecular structure of yukonite with As-O, As-Fe and As-Ca coordination numbers of 4, 3.24 and 4.17 respectively; but the molecular and structural nature of this phase still remains largely unfamiliar. In terms of environmental arsenic stability, Krause & Ettel (1989) found ~ 6-7 mg/L As released at pH 6.15 after 197 days; Swash & Monhemius (1994) found it to give approximately 2-4 mg/L As over the pH range 5-9 after 7 days, while recently Becze & Demopoulos (2007) found it to give an arsenic release of 1.16- 5.11 mg/L over a pH range of 7.5-8.8 after 24 hrs of TCLP-like testing. Extension of these tests (Becze et al., 2010) to longer period of time (> 450 days) was carried out both in gypsum-free and gypsum-saturated waters at various pH conditions where the environmental arsenic release was found to be significantly higher in gypsum-free (8.96 mg/L at pH 7, 47.8 mg/L at pH 8, and 6.3 mg/L at pH 9.5) tests compared to gypsum-saturated waters (0.75 mg/L at pH 7, 2 mg/L at pH 8, and 6.3 mg/L at pH 9.5.

Arseniosiderite is another closely related hydrated calcium ferric arsenate mineral relevant to these studies; it was first reported by Koenig (1889) as Mazapilite from the Jesus Maria Mine in Zacatecas, Mexico but was later shown by Foshag (1937) to be Arseniosiderite by comparison with a specimen from Mapimi in Durango, Mexico. This mineral is commonly found in other regions of the world such as Greece, Namibia, Spain, France, Austria, Germany, England, USA, Bolivia and Australia. Paktunc et al. (2003) reported the occurrence of arseniosiderite ($Ca_2Fe_3(AsO_4)_3O_2 \cdot 3H_2O$) and yukonite ($Ca_2Fe_3(AsO_4)_4(OH) \cdot 12H_2O$) in the Ketza River mine tailings produced from the operation of a former gold mine/mill in south central Yukon, Canada while in 2004 the same authors (Paktunc et al., 2004) found its presence in the gold ore from the Ketza River mine location. In a separate, study Fillipi et al. (2004) investigated the arsenic forms in contaminated soil mine tailings and waste dump profiles from the Mokrsko, Roudny and Kasperke Hory gold deposits in the Bohemian Massif found in the Czech Republic. It was reported based on chemical composition (EDX) and XRD analysis that arseniosiderite and pharmocosiderite [$KFe_4(AsO_4)_4(OH)_4 \cdot 6-7H_2O$] were the only phases present at the Mokrsko location in soils above the granodiorite bedrock. Later on Filippi et al. (2007) published an extensive report on the mineralogical speciation (via SEM-EDS/WDS, Electron microprobe, XRD and Raman spectroscopy) of the natural soils contaminated by arsenic found in the Mokrsko-west gold deposit (Czech Republic) where pharmacosiderite and arseniosiderite (in addition to Scorodite and jarosite) were identified as products of arsenopyrite (FeAsS) and/or pyrite(FeS_2) oxidation-weathering.

Several authors have commented on the properties of arseniosiderite (Larsen & Berman 1934; Chukhrov et al., 1958). Palache et al., (1951) and listed the crystal system of arseniosiderite as hexagonal or tetragonal, based on optical data. Moore & Ito (1974) reported arseniosiderite to have the following chemical formula [$Ca_3Fe_4(OH)_6(H_2O)_3(AsO_4)_4$] and to belong to a monoclinic crystal system [A2/a], eight molecular formula units per unit cell (Z = 8) and to be isostructural with robertsite [$Ca_3Mn_4(OH)_6(H_2O)_3(PO_4)_4$] and mitridatite [$Ca_3Fe_4(OH)_6(H_2O)_3(PO_4)_4$]. However, the precise formula of the mitradatite could not be unambiguously defined without detailed knowledge of its crystal structure. Therefore, Moore and Araki published (Moore & Araki, 1977a) an extensive crystallographic study on mitridatite (and as an extension on arseniosiderite) where they reported mitridatite ($Ca_6(H_2O)_6[Fe_9O_6(PO_4)_9] \cdot 3H_2O$) to possess a monoclinic structure [now with a space group, Aa] and Z = 4. Briefly, mitridatite is built up with compact sheets of [$Fe^{III}_9O_6(AsO_4)_9$]$^{-12}$

oriented parallel to the {100} plane (at the x ~ 1/4 and 3/4 crystallographic positions). These sheets are built up from octahedral edge sharing nonamers each defining trigonal rings which fuse at their trigonal corners to the edge midpoints of symmetry equivalent nonamers and are decorated above, below and in the plane by the PO_4 tetratehedra. In addition to these octahedral sheets, a thick assembly of $CaO_5(H_2O)_2$ polyhedra and water molecules occur as open sheets parallel to the {100} plane (at the x ~ 0 and ½ positions). The water molecules in these $CaO_5(H_2O)_2$ polyhedra define an OW-OW' edge, which is shared by two calcium atoms resulting in the formation of $Ca_2O_{10}(H_2O)_2$ dimers that contribute to the destruction of trigonal symmetry when they are placed on the octahedral sheets. This arrangement leads to the monoclinic cell symmetry of mitradatite. Arseniosiderite is an isotype of mitradatite, where arseniosiderite obtained from isomorphic replacement of $PO_4 \leftrightarrow AsO_4$. This replacement results in dilation along the "a" crystallographic direction because of the contribution in this direction by the larger arsenate tetrahedra is not constrained by the relatively rigid octahedra $\infty[Fe_9O_{33}]$ sheets. It is also worth noting that same year Moore & Araki (1977b) published a short synopsis where they stated that mitradatite (and as an extension arseniosiderite) had the same structure as previously mentioned but reported the space group to be A2/a. Finally, it is worth noting that to date no real crystallographic structure on arseniosiderite has ever been fully reported/published as in the case of isostructure mitradatite. Paktunc et al. (2003) reported via As K-edge EXAFS that the coordination number and bond lengths of the As-O and As-Fe units were essentially identical in yukonite and arseniosiderite, and that at the local structural level, the only difference between yukonite and arseniosiderite was the As-Ca coordination number (4.17 and 2.44, respectively). It should be noted that Paktunc et al. (2003) failed to reference any of the extensive work done by Moore & Ito (1974), and later Moore & Araki (1977a,1977b) on the structure of mitridatite (and as extension arseniosiderite) but still somehow managed to fit the EXAFS structure of arseniosiderite and came up with a As-Ca coordination number of 2.44. However, Moore & Araki (1977a) clearly stated that *"each Ca unit is coordinated to four phosphate/arsenate oxygens"* thus rendering the previous EXAFS results inaccurate. Moreover, in a subsequent publication, the same researchers (Paktunc et al., 2004) reported different As-Ca coordination numbers for the same phase (arseniosiderite), namely 2.44, 3.60 and 5.5, thus creating more uncertainty as to the true molecular identification of arseniosiderite and how it is distinct at the local molecular level to yukonite. The stability of arseniosiderite in terms of arsenic leachability was first investigated by Krause & Ettel (1989) who found that after 197 days in water, ~ 6-7 mg/L As was released into solution at pH 6.85. Swash & Monhemius (1994) also tested the arsenic release of arseniosiderite samples for 7 days and found them it to give release ~0.5-0.7 mg/L As over the pH range 5-9. No further arsenic stability tests for arseniosiderite exist to date but due to its crystalline nature they should perform similar or better than yukonite under the same conditions.

Since these Ca(II)-Fe(III)-AsO_4 phases (Yukonite and Arseniosiderite) were some of the most important arsenic phases found to occur in the Gold mining tailing operations around the world {Yukon territory,(Paktunc et al., 2003, 2004) and Nova Scotia, Canada (Walker et al. ,2005 and 2009)] as well in the Mokrsko-west gold deposits, Czech Republic (Filippi et al., 2004; Filippi et al., 2007; Drahota et al., 2009; Drahota & Filippi 2009)} and as a result of the fact that our experimental evidence showed that synthetic Ca(II)-Fe(III)-AsO_4 produced phases (in particular yukonite) resulted in excellent environmental arsenic retention properties (Becze & Demopoulos, 2007; Becze et al., 2010) under similar industrial conditions used in the Uranium industry disposal sites such as those found in Northern

Saskatchewan, Canada. Yet, in spite of all these studies and important relevance in academia as well as industry, the true molecular nature of these Ca(II)-Fe(III)-AsO$_4$ phases (yukonite and arseniosiderite) and how these phases were related was still not yet well understood. Therefore, a series of studies was conducted and published in this system via the use of several analytical structural and molecular techniques, of which vibrational spectroscopy was in particular an invaluable tool (Becze & Demopoulos, 2007; Becze et al., 2010; Gomez et al., 2010a).

2.2 Results and discussion

In the beginning of this work, no synthesis method for these Ca(II)-Fe(III)-AsO$_4$ phases had yet been developed in the academic or industrial literature, in spite of their great importance, unlike in the case of other industrially relevant arsenic phases (e.g. Scorodite). Some attempts were conducted and reported by P. M. Swash (1996) during his Ph.D. period under the supervision of A. J. Monhemius but no successful results were ever reported or published. These unsuccessful results were largely due to the fact that most of the syntheses conditions investigate by this research group were largely conducted in acidic media but as it is shown in the synthesis procedure developed by Dr. Levente Becze (Becze & Demopoulos, 2007; Gomez et al., 2010a; Becze et al., 2010), these phases are only stable at higher pH conditions (~ 8) and as such their formation also favours these environment. Once a reproducible and stable method of synthesis of these phases was developed we further investigated how the chemical distribution of the Ca, Fe, AsO$_4$, OH and H$_2$O were distributed among each of these Ca(II)-Fe(III)-AsO$_4$ phases.

Sample	Chemical Composition (w.t. %)			
	As	Fe	Ca	(H$_2$O + OH)
Synthetic	21.60 ± 0.55	26.52 ± 2.68	8.75 ± 0.63	14.8
Tagish Lake	23.08 ± 0.40	29.12 ± 0.58	6.90 ± 0.12	19.9
Romanech Arseniosiderite	23.86 ± 1.13	21.73 ± 1.01	10.35 ± 0.37	8.4

Table 1. Elemental analysis of several synthetic yukonite, natural yukonite and natural arseniosiderite. (Gomez et al., 2010a)

From our work (Becze & Demopoulos, 2007; Gomez et al., 2010a; Becze et al., 2010) on the synthetic and the natural specimens of these mineral phases, it was concluded that yukonite could accommodate a wide range of chemical composition (while still retaining the same phase) according to the general formula Ca$_2$Fe$_{3.5}$(AsO$_4$)$_3$(OH)$_4$·10xH$_2$O as shown in our studies (Table 1) and previous literature works. Arseniosiderite is very similar in terms of chemical composition to yukonite (Table 1) but as mentioned previously is a distinct mineral phase. Arseniosiderite has been indicated to be more crystalline counter part of yukonite (Garavelli et al., 2009) and generally has slightly more calcium content in both natural and synthetic (Paktunc et al., 2003, 2004; Garavelli et al., 2009; Gomez et al., 2010a) samples then yukonite. More interesting, it was noted that the chemical formula commonly used in the literature (Paktunc et al., 2003, 2004; Fillipi et al., 2004, 2007; Garavelli et al., 2009) Ca$_2$Fe$_3$(AsO$_4$)$_3$O$_2$·3H$_2$O had a peculiar form in that the use of peroxide types of molecular groups were used but also no hydroxyl groups were included in the formulas as in yukonite. Interestingly, most of these works where the formula was written in that fashion had no reference given to the earlier work of Moore and co-workers (Moore & Ito, 1974;

Moore & Araki, 1977) who determined the crystallographic structure of its phosphate isostructure-mitradatite. In both reports, the authors consistently reported that oxo-anions (O^{2-}) and/or hydroxyl units (OH^-) existed in the structure but certainty no peroxo anions ($O_2)^{2-}$. Moreover, even among the crystallographic reported works, the formulae also varied from $Ca_3Fe_4(OH)_6(H_2O)_3(XO4)_4$ in 1974 to $Ca_6(H_2O)_6[Fe_9O_6(XO_4)_9]$ •$3H_2O$ in 1977 and expressed in many alternative ways to describe its crystallographic arrangement but in no case was the use of peroxide expression used to describe its molecular formulae.

Therefore we decided to examine the vibrational structure of arseniosiderite to examine and verify if indeed these groups (O_2, OH, H_2O) were expressed in the molecular structure of this mineral phase. Peroxo or superoxo vibrational bands are known to be strongly Raman active and occur at 825-870 cm^{-1} or 1131-1580 cm^{-1} as in phases such as studtite ($UO_2(O_2)$ $4H_2O$) (Burns & Hughes, 2003; Bastians et al., 2004). The Raman spectra of arseniosiderite in our work and previous others (Filippi et al., 2007) did not appear to contain peroxo groups and as such it was determined that expressing the formulae as $Ca_2Fe_3(AsO_4)_3O_2$•$3H_2O$ was incorrect because it gives the wrong impression that peroxo units exist in arseniosiderite structure which is untrue. In addition, the question of whether hydroxyl groups occurred in the structure arose from the previous work reported by Moore and co-worker earlier work (1974) in which in the use of oxo-anions and hydroxyl anion were included in the formulae but in the later work (1977) only the use of oxo-anions was employed. Essentially, they arrived at this option by investigating bond length-bond strength variations by XRD. On the basis of this investigation they concluded that no hydroxyl (OH^-) anions occur in the structure and treated all anions as oxo-anion (O^{2-}) because they were all undersaturated (according to Baur's fourth rule $\Delta p = p_{donor} - p_{acceptor}$). Thermo Gravimetric Analysis could only show us how much $OH+H_2O$ existed in our sample but not could tell us if one or both definitely existed in the structure of this phase; more over previous X-ray crystallographic works had no additional information due to their lack of Hydrogen atom sensitivity (unlike neutron based diffraction).

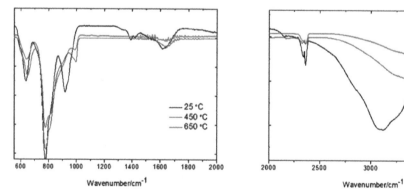

Fig. 1. ATR-IR spectra of arseniosiderite at 25 °C, 450 °C and 650 °C in the arsenate (left) and hydroxyl (right) stretching regions. (Gomez et al., 2010a)

Thus, in order to clarify this issue, we decided to conduct annealing/heating analysis (450 °C, 600 °C and 750 °C) of arseniosiderite similarly to that done by Garavelli et al. (2009) but this time using a molecular sensitive technique (ATR-IR) to determine whether there were indeed structural hydroxyl units present. From this analysis as shown above, it was confirmed that this structure have water molecules, hydroxyl groups that were found to still exist even after heating the sample to > 600 °C (Figure 1). Therefore, based on all this vibrational evidence our preference for a "new"(similar to work by Moore & Ito, 1974) expression formula $Ca_3Fe_5(AsO_4)_4(OH)_9 \cdot 8H_2O$ was adopted for this mineral rather than $Ca_2Fe_3(AsO_4)_3O_2 \cdot 3H_2O$. It should be noted that in the case of yukonite similar heating test were also conducted to verify that indeed both OH and H_2O groups are included in the structure, in agreement with previous works (Tyrrell & Graham, 1913; Jambor 1966; Ross & Post 1997; Pieczka et al., 1998; Paktunc et al., 2003, 2004; Nishikawa et al., 2006; Becze and Demopoulos 2007; Walker et al., 2009; Garavelli et al., 2009). Finally, in this work it was observed that for arseniosiderite, and additional arsenate IR and Raman active vibrations occurred ~ 900 cm^{-1} indicative of protonated arsenate groups. Another such possibility for the existence of additional vibrational bands at such higher wavenumbers is the possibility of more than one type of crystallographic arsenate group which would add extra contributions to the arsenate vibrational region of the spectra. However, this latter possibility is unlikely from inspection of the lack of splitting to the v_3 arsenate modes which should be additionally splitted if more than one distinct crystallographic arsenate molecule occurs in the crystal structure (for example in Ferric Arsenate sub-hydrate in section 3.2); more over from the crystallographic data of Moore and co-workers (Moore & Ito, 1974; Moore & Araki, 1977a) only one distinct arsenic position is described in its structure. Therefore, the presence of this band in the Raman and IR spectra of arseniosiderite clearly indicates that not only do arsenate groups occur in the structure of this phase but also protonated arsenate groups, something that no previous works on this phase have ever mentioned or notice. Although one brief Raman report was published (Fillipi et al., 2007), the focus of this study was not to investigate the molecular structure of this phase and as such the use of Raman spectroscopy was employed only as a fingerprint tool. However in our case, as a result of the combination of vibrational techniques (IR and Raman) along with a detailed crystallographic knowledge of the structure, new molecular information was obtained.

As can be observed from the brief analysis above (the chemical composition coupled with the vibrational work) it is apparent that a molecular formulae/composition was obtained in both phases (yukonite and arseniosiderite) but their molecular arrangement and local structure was still unknown. This was largely as a result of the fact that although arseniosiderite is a crystalline material, its actual crystal structure has not yet been actually determined (only inferences from its isostructures such as mitradatite have been actually reported).

		Coordination Number	Bond Length (Å)
Arseniosiderite	As-O	4.25	1.71
	As-Fe	3.66	3.29
	As-Ca	2.44	4.24
Yukonite	As-O	4.00	1.71
	As-Fe	3.24	3.28
	As-Ca	4.17	4.21

Table 2. Arseniosiderite and Yukonite's first three shell (As-O, As-Fe and As-Ca) coordination numbers and bond lengths based on fitting of As K-edge EXAFS data. (Paktunc et al., 2003, 2004)

Yukonite is a semi-crystalline unordered type of phase whose structure at the macro or nano level also still remains unknown. From previous literature data (Paktunc et al., 2003, 2004) on As K-edge EXAFS measurements of yukonite and arseniosiderite, the local As-Ca coordination of these two closely related phases which is the only significant difference at the local coordination level to distinguish these two phases arose to question. Moreover, the reports of two distinct As-Ca coordination numbers for arseniosiderite by the same author on subsequent years (Paktunc et al., 2003, 2004) gave rise to even more confusion on this subject (Table 2). Although vibrational spectroscopy can offer us qualitative insight into the coordination of molecules in a crystal structure, most often this occurs as an average of all the molecules. Moreover, its probing range of the structure resembles that of a long range probe due to the size of the wavelength radiation used in these techniques (IR and Raman) and as such, element specific local coordination information is not possible to attain with vibrational spectroscopy. Based on this latter fact and as a result of the contradictory results from previous works (Paktunc et al., 2003, 2004) we decided to conduct our own XAS analysis at the As K-edge and at the Fe and Ca L-edges. From the As K-edge and Fe L-edge X-ray Absorption Spectroscopic analysis (refer to Gomez et al., 2010a) it was determined that in fact the local molecular environment of arsenic and iron of yukonite and arseniosiderite was almost identical (in terms of bond lengths and coordination numbers) to each other, in agreement with that of previous reports.

In complicated systems such as arseniosiderite and yukonite, the Ca L-edge XAS (in particular the XANES region) provides a sensitive probe to determine the local coordination structure at the selected atomic site within 4-5A° (Borg et al., 1992) from the core atom (in our case Calcium). Therefore it can provide specific information on the local character of the $3d^0$ unoccupied states via the $2p^5 3d^n$ excited electronic state (de Groot et al., 1990; de Groot, 2005). The multi-peaks (a_1, a_2, b_1, b_2) observed in the Ca L-edge XANES spectrum have been attributed to the crystal field arising from the symmetry of the atoms surrounding the Ca^{2+} ions leading to atomic and charge transfer multiplet effects. These effects have been shown to add extra features and redistribute the intensity over all the lines in the XANES spectrum (Naftel et al., 2001; de Groot, 2005). Moreover, the splitting of these features can be related nonlinearly to the value of the crystal field splitting parameter, coordination and site symmetry around the calcium atoms as well as spin-orbit splitting at the 2p level. Therefore in our studies, the use of the Ca L-edge XAS was employed to determine if in fact the As-Ca coordination was distinct in yukonite and arseniosiderite as previously reported. If indeed the calcium local As-Ca coordination was different in yukonite and arseniosiderite, then the Ca L-edge XANES spectra that should be observed should also be distinct (Naftel et al., 2001; de Groot, 2005; Fleet & Liu, 2009). As may be seen from the Ca L-edge data (Figure 2), the calcium exhibits the exact same spectra indicating that the same type of calcium coordination and crystal field is observed both yukonite and arseniosiderite. Thus, from the Ca L-edge XANES spectra (Figure 2) it may be inferred that the electronic and local calcium coordination state (As–Ca) and thus coordination number in yukonite is very similar or exactly the same as in arseniosiderite, in contrast to previously reported EXAFS data (Paktunc et al., 2003, 2004). Therefore based on this information and that of previous works on arseniosiderite and yukonite (Moore and Araki, 1977; Paktunc et al., 2003, 2004) it was concluded that "each Ca unit is coordinated to four phosphate/arsenate oxygen's" in both structures (Moore & Araki, 1977).

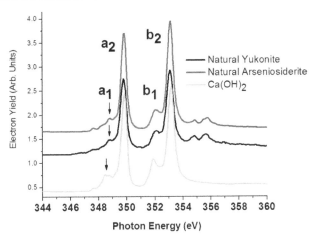

Fig. 2. Ca L-edge XANES of yukonite and arseniosiderite. The spectrum of Ca (OH)₂ is also shown for reference as a Ca^{+2} standard in an octahedral crystal field. (Gomez et al., 2010a)

From these above findings it can be inferred that from a chemical point of view and local coordination structural aspect (XAS results), it was apparent that these structures were only different from the fact that one was a crystalline (arseniosiderite) counter part of the other (yukonite) but aside from this both appeared to be identical in every respect.

SSo, the questioned remained: what exactly makes these two distinct mineral phases not related structurally (aside from crystallinity) or molecularly?

By further investigating the nano structure of these closely related minerals, it was observed that although yukonite was a semi-crystalline composed of glass like micro aggregates, at the nano-scale under the TEM, yukonite was remarkably found to be composed of nano-domained structures which actually showed order at the nanometer level. In case of arseniosiderite, as expected, it was composed of micro ordered domains and showed electron diffraction typical of single crystal domains (Figure 3). Raman spectroscopic measurements indicated that even yukonite's and arseniosiderite's Raman structure of arsenate groups and lattice vibrational modes appeared similar for both phases except for the appearance of the IR and Raman active protonated arsenate bands (HAsO₄) for arseniosiderite. The Raman vibrational bands were found to be broader in the case of yukonite (synthetic and natural) versus the sharper bands observed for arseniosiderite, confirming that the AsO₄ molecules in arseniosiderite were in higher long range order than in yukonite. This of course is in agreement with data and conclusions observed from the XRD and TEM data. Looking at the ATR-IR spectra of these two such closely related phases it was apparent that although their arsenate and Fe-OH/OH₂ structure was similar in nature, there was an additional feature that only IR spectroscopy allowed us to observe.

The infrared hydroxyl region (3000-4000 cm^{-1}) has the advantage of giving us insights into the intermolecular hydrogen bond nature of minerals (Sumin de Portilla, 1974; Hawthorne, 1976). For example in Scorodite, where two distinct types of intra-molecular hydrogen bonding occurs between the crystal water molecules of metal octahedra and the arsenate groups of the form MO – H---OAs; it is possible to relate each specific water (OH) stretching to a particular vibrational band of the two distinct crystallographic water sites (Hawthorne, 1976; Gomez et al., 2010b). The relative strength of H-bonding interactions between the two

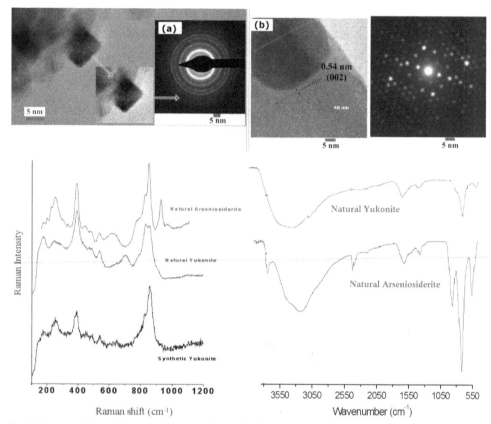

Fig. 3. Transmission Electron Micrograph and Selected Area Diffraction of yukonite (left-top) and arseniosiderite (right-top). Raman (left-bottom) and ATR-IR (right-bottom) spectra of yukonite and arseniosiderite is also shown below. (Gomez et al., 2010a)

distinct water molecules may be inferred by observing their relative vibrational positions and calculating their H-bond distances using Libowitzky's correlation functions (Libowitsky, 1999). In general, OH stretches at lower wavenumbers are stronger in bond strength (smaller H-bond bond distances) then those at higher wavenumbers. In the case of arseniosiderite two hydroxyl stretches (3100, 3576 cm-1) are observed, almost identical to that found in Scorodite (Hawthorne, 1976; Gomez et al., 2010b, 2011b). These two OH stretches in arseniosiderite are indicative of two distinct types of H-bonding environments (one strong and one weak); yukonite on the other hand exhibits only a diffuse band (3111-3215 cm-1) indicative of a disordered type of H-bonding, typical of its glassy disordered nature. This latter distinct feature of ordering at the molecular level via H-bonding is a characteristic of each phase; this latter fact gave for the first time, a molecularly based difference between these two closely related phases (yukonite and arseniosiderite) that did not derive from their crystallinity. Based on the Ca 2p, Fe 2p and As-K edge XANES, it is apparent that yukonite exhibited the same local calcium, iron and arsenic local structural environment and units encountered in arseniosiderite but with less long range order in the crystal lattice domains (as observed via X-ray and TEM). Therefore, the lower long range order in yukonite is proposed to arise from lack of developed hydrogen bonding environment (observed via ATR-IR) giving rise to poor long range order and is then expressed physically as nano-domain size randomly oriented lattices. In contrast, arseniosiderite displayed a well ordered H-bonding system in its structure giving rise to long range order and a crystalline structure which was physically manifested via micro-domain single crystal lattices (TEM). Therefore, based on these results, a molecular and structural link between these two otherwise similar phases (yukonite and arseniosiderite) was proven via the use of vibrational technique coupled with other analysis techniques.

3. The hydrothermal Fe(III)-AsO$_4$-SO$_4$ synthetic system and their relation to industrially produced phases

3.1 Introduction

The ferric arsenate waste solids produced from the hydrometallurgical chemical processing of mineral feedstock's from various metal ores (Au, Co, Mo, Zn, U) can be split into two groups depending on their degree of crystallinity (Riveros et al., 2001). At ambient temperature and pressures, poorly crystalline Fe(III)-As(V) solids (Krause & Ettel, 1989; Langmuir et al., 1999; Jia & Demopoulos, 2005, 2007) are produced by co-precipitation, which consist of ferric arsenate and arsenate-adsorbed onto ferrihydrite. (Langmuir et al., 1999; Jia & Demopoulos, 2005, 2007) These co-precipitates are produced from high Fe (III) to As(V) molar ratio solutions (typically >3) by lime neutralization (Riveros et al., 2001). Today, this method is still considered to be the most suitable method to treat low arsenic containing process effluent solutions.

In the case of arsenic-rich and iron deficient solutions, crystalline phases such as Scorodite (FeAsO$_4 \cdot 2H_2O$) can be produced. This can be done for example at elevated temperatures, near the boiling point of water (80-95 °C) and under controlled supersaturated conditions. (Singhania et al., 2005; Demopoulos, 2005; Fujita et al., 2008) Crystalline Scorodite is at least 100 times less soluble than its amorphous counterpart (FeAsO4 · xH2O[am]) (Krause and Ettel, 1989; Langmuir et al., 2006; Bluteau & Demopoulos, 2007) and given its high arsenic content (in comparison to the Fe (III)-As(V) co-precipitates as mentioned above) has been advocated for the fixation of arsenic-rich wastes. (Filippou & Demopoulos 1997; Fujita et al., 2008) At even higher temperatures (>100 °C, the hydrothermal precipitation range) during autoclave hydrothermal processing of copper (Berezowsky et al., 1999) and gold (Dymov et al., 2004) sulphide feedstock's, other crystalline phases than Scorodite are reported to form some of which exhibit equal or better stability than Scorodite (Swash & Monhemius, 1994; Dutrizac & Jambor, 2007, Gomez et al., 2011a).

Swash & Monhemius (1994) were the first to report on the precipitation and characterization of Fe(III)-AsO$_4$ compounds from sulfate solutions under industrial like autoclave processing conditions. In their work, four distinct crystalline phases were found to form which were Scorodite, FeAsO$_4$ ·2H$_2$O; Basic Ferric Sulfate, FeOHSO$_4$ (BFS); Type1", Fe$_2$(HAsO$_4$)$_3$ ·zH$_2$O with z=4; and "Type 2", Fe$_4$(AsO$_4$)$_3$(OH)$_x$(SO4)$_y$ with x+2y=3. The formation of these phases was correlated to two formation variables: temperature and Fe(III) to As(V) molar ratio using a fixed retention time. "Type 2" (0.34 mg/L As) was found to meet the TCLP leachability criterion exhibiting similar behavior with Scorodite (0.8 mg/L As). A decade later, Dutrizac & Jambor (2007) published an extensive experimental program involving the precipitation of Fe (III)-AsO$_4$-SO$_4$ phases. In their program, the effects of time, initial acidity and variable Fe(III), As(V) concentrations were considered. The characterization results via elemental analysis and XRD identified two new phases in addition to Scorodite (FeAsO$_4$·2H$_2$O) and basic ferric sulfate (BFS: FeOHSO$_4$). The two new phases, were labeled as "Phase 3, Fe(AsO$_4$)$_x$(SO$_4$)$_y$(OH)$_v$(H$_2$O)$_w$ where x+y=1 and v+w=1" and "Phase 4, FeAsO$_4$·3/4H$_2$O. Short term (40 h) leachability tests yielded 0.1 mg/L As for Phase 3 and 1-3 mg/L As for Phase 4 indicating that Phase 3 might be an acceptable carrier for the disposal of arsenic. In these studies, the new phases and results were found to be completely different than in the studies of Swash & Monhemius (1994).

Therefore since the true characteristics, conditions/mechanisms of formation, and long term arsenic environmental stability had not been unequivocally established or determined; a series of synthetic, characterization and environmental arsenic stability tests were conducted and published via the use of several techniques with vibrational spectroscopy being particularly important as will be shown below (Gomez et al., 2010b, 2011a, 2011b).

3.2 Discussions and results

The first portion of this work was to synthesize and chemically analyze these Fe(III)-AsO$_4$-SO$_4$ phases under the conditions conducted by previous studies (Swash & Monhemius, 1994; Dutrizac & Jambor, 2007) but also using our own distinct reacting conditions and reaction variables (Gomez et al., 2011a). From our work and comparison to those previously studied, it became apparent that in spite of the various reaction conditions and variables that each study employed, there was a general agreement in all studies which indicated the formation of similar separate phases from a chemical and structural perspective. However, most of the previous works on this system (Swash and Monhemius, 1994; Dutrizac and Jambor, 2007) have largely based their attention to chemical and structural type of analysis (ICP-OES, XRD, and SEM) and no real consideration of molecular analysis was ever undertaken. This lack of molecular analysis on the previous works part is due to the fact that these bulk-structural types of techniques are the most frequently encountered in these research areas/fields (hydrometallurgical process engineering). It should be pointed out that although one work was published on the IR spectra of some of these phases, little or to no molecular information was ever inferred from the published works available nor any complementary Raman analysis was ever conducted to analyze these types of materials (Swash, 1996; Ugarte & Monhemius, 1992).

Investigation	McGill			§Swash & Monhemius 1994			§Dutrizac & Jambor 2007		
Weight %	Fe	As	S	Fe	As	S	Fe	As	S
Scorodite	23-25	28-33	0.2-2	20-24	27-37	≤1.3	23-25	28-30	0.9-2.5
FAsH/Type1/Phase4	25-28	32-42	0-0.4	21-23	30-40	<0.33	26-27	36-38	0.2-0.4
BFAS/Type2/Phase3	23-32	14-27	3.2-9	27-38	17-28	≤4.3	29-30	13-23	8-13

§ Note: AsO$_4$ and SO$_4$ values were converted to As and S values for comparison.

Table 3. Elemental composition of the three main phases of interest found for our work (Gomez et al. 2011a) in comparison to the other previous (Swash & Monhemius, 1994) and "new" reported phases (Dutrizac & Jambor, 2007).

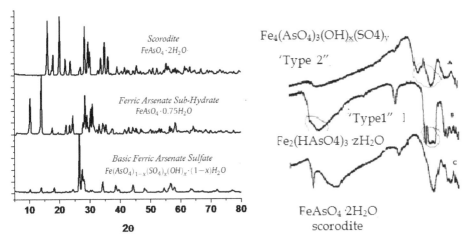

Fig. 4. Powder XRD (left) of the three main phases produced in our studies (Scorodite, Ferric Arsenate sub-Hydrate and Basic Ferric Arsenate Sulfate) (Gomez et al., 2010b, 2011a). The FTIR spectra of scorodite, Type 1 (same as FAsH) and Type 2 (same as BFAS) using transmission collection mode and KBr pellets(Swash, 1996; Ugarte and Monhemius,1992).

This lack of molecular information from previous works (Swash, 1996; Ugarte and Monhemius,1992) was a result of the fact that these previous authors did not bother to investigate which collection mode of Infrared spectroscopy (Transmission, Diffuse Reflectance, Attenuated Total Reflectance or Photo acoustic) would be best to analyze the vibrational structure of these phases and as such they decided to go with the use of the Transmission mode (KBr dilution). As can be observed in Figure 4 above, the IR spectra of the phases relevant to these studies lacked the fine structure and resolution necessary to extract any vibrational and molecular information for the various iron arsenate phases produced.

It is noted here that the mode of collection is very important to consider as different samples give rise to better signals for detection in different IR collection modes. In our work (Gomez et al., 2010b, 2011a, 2011b) care was taken to analyze these opaque types of powders in both transmission as well was reflection modes; after careful evaluation of the spectra for each of these phases it was decide that the Attenuated Total Reflectance (ATR) mode would give us enough of an energy range as well as the required spectra resolution to be able to extract useful molecular information of these phases. The ATR mode is especially suited for opaque powders, thin films and solutions, while the transmission mode often tends to 'smear' vibrations due to the less light transmission or detection in opaque solids, (i.e. poorer signal is collected by the detector). Indeed, comparison between the previous IR transmission data (Swash, 1996; Ugarte and Monhemius, 1992) and our work (via ATR) for the same type of phases (Figure 4 and 5) we can clearly see that not only is there more vibrational structure in the regions of interest (arsenate and sulfate v_3 modes) but we also get the advantage of getting much more additional structure in the hydroxyl region, something that will become important in the molecular analysis of these phases.

Once having the proper vibrational technique to observe as much of the vibrational structure as possible, it was imperative to determine structurally and molecularly if the phases produced in our studies and previous works (Swash & Monhemius, 1994; Dutrizac &

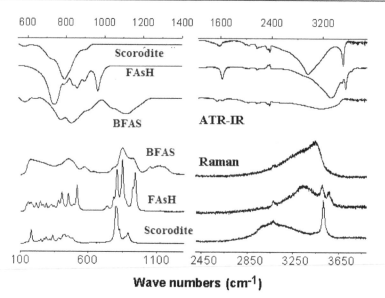

Wave numbers (cm⁻¹)

Fig. 5. ATR-IR (top) and Raman of the three main arsenate phases produced in our studies. These were Scorodite, Ferric Arsenate sub-Hydrate and Basic Ferric Arsenate Sulfate. Note the much richer IR structure observed in the arsenate and hydroxyl stretching region for these phases (e.g. Type 1 = FAsH) in comparison to that of the Transmission mode used in Figure 4 above.

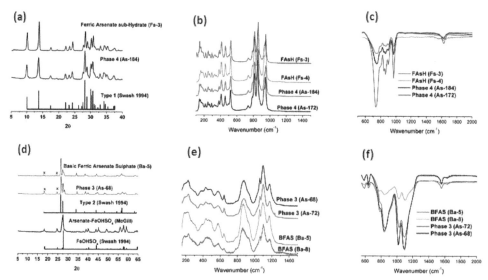

Fig. 6. Powder XRD (left), Raman (middle) and ATR-IR (right) of previous and new reported phases (Type 2, Phase 3, Type 1 and Phase 4) in comparison to the products produced in our studies (Ferric Arsenate Sub-Hydrate and Basic Ferric Arsenate Sulfate). (Gomez et al 2011a)

Jambor 2007) were all the same, especially in the case of the new claimed phases (Phase 3 and Phase 4) found by Dutrizac and Jambor (2007). From a chemical perspective these phases appeared to be the same in all studies (Table 3) but in the previous published works (Dutrizac & Jambor, 2007) it was claimed that these phases (namely Phase 3 and Phase 4) were unlike those produced by previous studies (Swash & Monhemius, 1994; Ugarte & Monhemius 1992).

As can be seen from the above (Figure 6), it is clear that at the structural level these new claimed phases (Phase 3 and Phase 4) were exactly identical to those produced in our studies (Gomez et al., 2011a) but more importantly those produced in previous studies (Swash & Monhemius, 1994). Not only was this confirmed at the structural level but more importantly we showed that at the molecular level the arsenate, sulfate and hydroxyl molecules of interest all exhibited the same fingerprint signal but more importantly the same molecular group symmetry exhibited in their crystal structures. Once we identified that all the phases produced under the particular reaction conditions were all the same phase(s), our detailed vibrational and factor group analysis on these phases was conducted (Gomez et al., 2010b).

In the case of Scorodite ($FeAsO_4 \cdot 2H_2O$) our structural and molecular results agreed well with that of the previous studies but the use of vibrational spectroscopy allowed us to spectrocopically observe the molecular incorporation of $SO_4 \leftrightarrow AsO_4$ substitution into the Scorodite structure, something which was postulated previosly from elemental analysis but had never been shown in detailed fashion via the use of vibrational spectroscopy until our work (Gomez et al., 2010b, 2011a, 2011b); more over the combination of elemental and molecular analysis allowed us to reformulate the chemical formula of Scorodite in a more appropiate manner as $Fe(AsO_4)_{1-0.67x}(SO_4)_x \cdot 2H_2O$.

A similar case occured in the case of the Ferric Arsenate sub-Hydrate (FAsH) = Phase 4 = Type 1, where it was observed that all these had a similar Fe, AsO_4 and OH/H_2O chemical composition, more over the crystal structure as shown via XRD was also the same. However, a contradiction was found yet again in this sytem as the chemical formula proposed for "Type1", $Fe_2(HAsO_4)_3 \cdot zH_2O$ by Swash & Monhemius (1994) was quite distinct from our work in which we proposed FAsH to be of the form $FeAsO_4 \cdot 3/4H_2O$ based on the crystallographic analysis of this phases using literature data (Jakeman et al., 1991) and XRD simulations with CaRine and Match diffraction software's (Gomez et al., 2011a). Dutrizac and Jambor (2007) indicated that the XRD pattern of their Phase 4 did not match that of any of the ferric arsenate structures in the International Centre for Diffraction Database, but indicated that its crystal structure likely was in the form as that reported by Jakeman et al. (1991) but no XRD simulations were done to verify these results. Therefore to clarify this dilemma of whether this phase actually contained AsO_4 or $HAsO_4$ or both in its structure, the use of molecular sensitive techniques such as X-ray Absorption Spectroscopy (NEXAFS and EXAFS) and vibrational spectroscopy (ATR-IR and Raman) was employed.

It has been well documented in the literature that the use of XAS (NEXAFS and EXAFS) can be routinely used to determine and distinguish whether there exist AsO_4 or $HAsO_4$ in a structure (Fernandez-Martinez et al., 2008; Guan et al., 2008) as a result of the difference and bond distances they exhibit and their change in symmetry (Myneni et al., 1998). For example, the As-OH bonds are elongated compared to the As-O bonds which is typical for protonated AsO_4 tetrahedra (Myneni et al., 1998) and leads to strongly distorted polyhedra. The mean As-O bond lengths in all AsO_4 tetrahedra of the three isotypic compounds range from 1.686 - 1.697 Å, which are longer than the average bond length for non-protonated tetrahedral (1.682 Å). More over as described in the section above, the XAS spectra is particularly sensitive to changes in coordination state due to its close relation to the crystal field which affects the empty density of states that is expressed in the XAS spectra. Therefore in our case, we expected to have these features be exhibited in the As K-edge NEXAFS or EXAFS data (Figure 7). However upon comparison of a known of Scorodite (which only has AsO_4) with our FAsH phase, we observed little to no difference in either phase nor any particular contributions from the $HAsO_4$ groups. Therefore, as a result of this a more careful treatment (via the use of factor group analysis) of the vibrational data of the FAsH (and Phase 4) was conducted to determine the types of arsenate groups found in its structure. From the vibrational literature data (Frost et al., 2006; Schwendtner & Kolitsch, 2007) of compounds that include AsO_4 and $HAsO_4$, it is known that protonated arsenate groups, these usually show stretching vibrations at higher wavenumber from 950 to 700 cm^{-1}, while that of unprotonated groups such as Scorodite shows a band ~ 700-840 cm^{-1}. In the case of FAsH=Type 1=Phase 4 both the IR and Raman spectra (Gomez et al., 2010b) display a strong band 940-960 cm^{-1}; this band is observed even in the poorly resolved IR data of previous works (Swash, 1996; Ugarte & Monhemius 1992) who suggested the presence of $HAsO_4$ versus AsO_4 groups in this phase. However, as noted previously the crystal structure of this phase was solved before all these studies began (Swash & Monhemius, 1994 and Dutrizac & Jambor, 2007) by Jakeman et al. (1991) who indicated that indeed this phase was of a triclinic form with a space group C_i and $Z = 4$. More importantly, the crystallographic results determined that this phase had two distinct crystallographic arsenate atoms which lie on two different C_1 crystallographic symmetries; similarly, the H_2O molecules all occupy C_1 sites and more importantly there also exist two types of H_2O molecules in the structure (one covalently bound to the iron octahedra and one electrostatically bound and placed along the channels of the structure). Therefore, with knowledge of the crystallographic arrangement along with the corresponding factor group analysis of the arsenate groups (Gomez et al., 2010b and 2011b) of this phase, it was easy to determine that in fact those vibrations observed at ~940-960 cm^{-1} were not from protonated arsenate groups but rather were part of the v_3 (AsO_4) stretches and in fact the six vibrations observed in the Raman and IR from ~700-960 cm^{-1} was actually a combination of the $v_3(AsO_4)$ and $v_1(AsO_4)$ coming from the two crystallographically distinct arsenate molecules in the structure (Figure 5). Therefore based on the vibrational analysis and the crystallographic

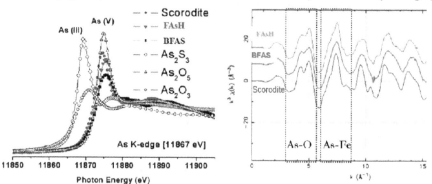

Fig. 7. As K-edge XANES (left) and EXAFS (right) data for Scorodite, Ferric Arsenate sub-Hydrate and Basic Ferric Arsenate Sulfate.

results from experimental and simulated data (Gomez et al., 2011a), it was determined that in fact the molecular formula for FAsH=Type 1= Phase 4 should not be of the form"$Fe_2(HAsO_4)_3 \cdot zH_2O$" but rather "$FeAsO_4 \cdot 3/4H_2O$" where the exact amount of water was determined via TGA analysis. Finally, it is interesting to note that in the hydroxyl region of FAsH, there is four distinct vibrations that occur in both the IR and Raman spectra which can be attributed to the two distinct types of water molecules in the crystal structure each which exhibit two distinct types of H-bonding {as in the case of Scorodite (Gomez et al., 2010b, 2011b)}.

In the case of Basic Ferric Arsenate Sulfate (BFAS) as mentioned above, no crystal structure existed to date (until recently in our later work; Gomez et al., 2011d) and as a result a detailed group analysis of the vibrational structure could not be conducted (Figure 8). Rather, the use of vibrational spectroscopy was instead used to monitor the $SO_4 \leftrightarrow AsO_4$ substitution into the structure and to infer molecular symmetry information for our future crystallographic refinement (Gomez et al., 2011d). In particular it was observed that as the variation of the $SO_4 \leftrightarrow AsO_4$ substitution occured in the structure of BFAS, the vibrational structure in the Raman spectra was qualitatively sensitive to the concentration of the groups in the structure but less so to the environment/symetry of the molecules. For example in the case where the sulfate (or arsenate) concentration increased, its Raman band appeared higher in relative intensity. Conversely, the ATR-IR vibrational structure was more sensitive to the environment/symetry of the molecular group of interest and less so to the amount of each group in the structure. In particular, it was observed that the solid solution of $SO_4 \leftrightarrow AsO_4$ not only changed the amount of each group incorporated into the BFAS structure but also actually changed the symmetry of the molecular group expressed in the vibrational spectra and more importantly showed us that these groups could occupy the same type of crystallographic sites in the crystal structure, something that would become important later in the determination of its crystal structure(Gomez et al., 2011d).

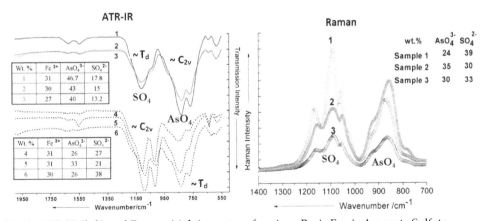

Fig. 8. ATR-IR (left) and Raman (right) spectra of various Basic Ferric Arsenate Sulfate products that contain various amounts of arsenate and sulfate in the solid determined via ICP-AES analysis. (Gomez et al. 2010b)

In terms of industrially produced samples, our work analyzed two industrially manufactured samples from the gold and copper industries (Gomez et al., 2010b, 2011c). In general for industrial practices XRD is the most common form of analysis employed due to the fact that is the easiest and most widely available tool with a large enough database for users to simply "click buttons" to get their desired solutions. This approach is often good enough to yield some reasonable results if a priori knowledge of species is assumed based on the matching of chemical elements from elemental analysis; however, this approach fails when the concetration of the phases is below the detection limit of lab based XRD (~ 4-5 w.t.%) or when the sample is not perfecly crystalline in nature. Indeed in our studies (Gomez et al., 2010b, 2011c) both scenerios were encountered in the industrial samples analyzed (Dymov et al., 2004; Defreyne et al., 2009; Mayhew et al., 2010; Bruce at al., 2011).

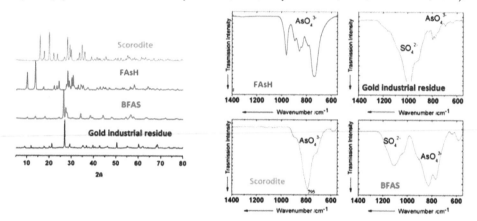

Fig. 9. Powder XRD (left) and ATR-IR (right) of an arsenic containing industrial residue produced from the gold industry compared against relevant synthetic iron-arsenate phases that form under similar conditions. (Gomez et al. 2010b)

For example, in the case of the gold produced residue (Figure 9), although the sample was crystalline in nature, the amount of arsenic (as AsO_4) in the solid was ~ 1 w.t. % and as a result trying to identify the type of arsenic form produced in the residue was almost impossible with XRD. The use of Raman spectroscopy on this industrial sample gave rise to no Raman active bands by means of four laser wavelengths (488, 514, 632 and 785 nm) and two different spectrometers (bulk and micro). However, upon the use of the ATR-IR technique which can achieve much lower detection limits as well as has the ability to tune to the molecular groups of interest groups of interest while rejecting others if used appropriately (see below); the form of arsenate appeared more clearly as may be seen in the figure above, where in general it only resembled the BFAS type of form and not the other two related ferric arsenate phases.

In the case of the copper produced industrial sample (Figure 10), due to the lower temperatures employed (Bruce at al., 2011) during the copper process (~ 150 °C), the crystallinity of the industrial sample was found to be much lower in comparison to the gold residue products (~ 230 °C). Moreover, the arsenic (as AsO_4) content found in this industrial residue (~ 1 w.t. %) was again much lower than the detection limits allowed by lab based XRD. Therefore, the lower crystallinity and low content of arsenic in this phase made the positive identification of arsenic phases much less accurate even with Rietveld type of analysis or simple database "click" search matches. As a result, vibrational spectroscopy was employed to investigate the arsenic phase found in this industrial residue. In the case of

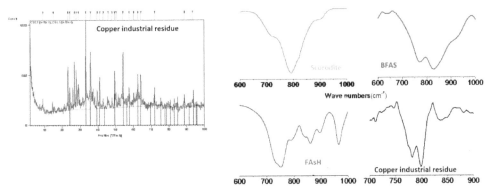

Fig. 10. Powder XRD (left) and ATR-IR (right) of an arsenic containing industrial residue produced from the copper industry. The ATR-IR spectra also shows a comparison of the residue against synthetic iron-arsenate phases which are known to form under similar conditions. (Gomez 2011c)

the Raman spectroscopy (data now show here but refer to Gomez et al., 2011c), vibrational structure was obtained unlike the other case but was found to be largely dominated by the other expected species found in the sample as a result of the chemical process it undergoes (Defreyne et al., 2009; Mayhew et al., 2010; Bruce at al., 2011). For example, it is expected to have both hematite and elemental sulfur as the major constituent of this produced residue; however, upon the analysis of this product with Raman spectroscopy, only hematite could be observed and not in any way elemental sulfur nor any arsenate type of phase. It was only after the elemental sulfur was removed via the chemical method describe in some of our earlier work on other industrial samples (Becze et al., 2009) that the elemental sulfur was extracted and re-crystallized, then Raman spectra was collected to ensure that indeed it was elemental sulfur. Before or after extraction of the elemental sulfur no Raman active arsenate phases were detected. Therefore, the use of ATR-IR spectroscopy was employed again to observe the type of arsenate phase found in the residue, this was done as it offers the advantage of being transparent to the other major species (hematite and elemental) expected to be found in the residue (both which exhibit distinct vibrational signal below 500 cm^{-1}) while the distinct arsenate mode for our phases of interest occurred at 700-900 cm^{-1}. It was found by comparison to our other vibrational data that the arsenate species found in the copper industrial residue again resembled that of our BFAS phase.

It is worth for the reader to also note that in our work (Gomez et al., 2010b, 2011a, 2011c) only three phases (and intermixtures of these) could be detected, namely Scorodite, Ferric Arsenate Sub-hydrate/FAsH, Basic Ferric Arsenate Sulfate using the lab based techniques (XRD, ATR-IR, Raman) employed in these studies. Interestingly, Basic Ferric Sulfate (FeOHSO$_4$) was a species detected in previous works formed at high (200-225 °C) and higher Fe/As ratios (~4) but no clear evidence was obtained from our work using the lab based methods which included vibrational spectroscopy. This was partly as a result of the fact the crystal structure of BFAS had not yet been determined (until recently, Gomez et al., 2011d) and the fact that in order to detect confidently such mixture of species higher resolution synchrotron based XRD and Rietveld refinement was necessary to verify these results, something that previous studies never considered or employed.

4. The TiO₂ binding mechanism with the N719 photo-sensitizer for dye-sensitized solar cell applications

4.1 Introduction

As a result of our large demands for energy consumption as our world population increases, the need for alternative energy sources are currently intensively investigated, in particular Dye-sensitized solar cells (DSSCs) have been given major attention due to their application in solar energy conversion and photovoltaic systems (Gratzel, 2001; Kroon et al.,2007; Murakoshi et al., 1995; Falaras, 1998; Finnie et al., 1998; Nazeeruddin et al., 2003; Leon et al., 2006; Kay & Gratzel, 2002; Shklover et al., 1998; Hagfelt, 2010). In particular, the anatase TiO₂ semiconductor/Ru complex [*cis* - (2,2'- bipyridyl - 4,4'- dicarboxylate)₂ (NCS)₂ ruthenium(II): N719] interface has been studied to understand the sensitization event via the use of spectroscopic [vibrational(Murakoshi et al., 1995; Falaras, 1998; Finnie et al., 1998; Nazeeruddin et al., 2003; Leon et al., 2006; Hagfelt, 2010) and X-ray absorption(Kay & Gratzel, 2002; Shklover et al., 1998)] and computational studies (De Angelis et al., 2007a, 2007b).

In spite of this research efforts, the binding mechanism (Figure 11) has been debated for decades and remains of importance as it relates to electron transfer and ultimately to the performance of the dye-sensitized photoanode. For example, strong bonding interactions such as covalent bonding (owing to the strong electronic coupling between the semiconductor d-states of TiO₂ and the N719 dye molecular orbital) lead to fast electron injection processes and impact the electronic performance of the DSSC (Murakoshi et al., 1995; Wang and Lin, 2010; Bazzan et al., 2011). Moreover, the presence of non (H-bonded) or single (monodentate-ester type) bonds versus two covalent bonds between the TiO₂ and N719 molecule are likely to affect the type and amount of electron transfer that occurs during the photo-injection process. For example, two electron bridges in the binding (as bidentate) is likely to favor more electron transfer in the N719-TiO₂ system versus monodentate type (or electrostatically) type of coordination where only 1 electron bridge occurs (or where no direct link occurs).

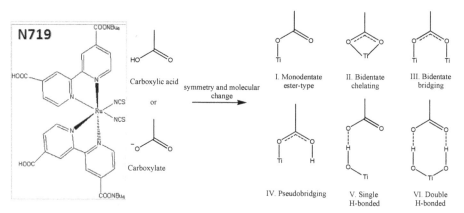

Fig. 11. The N719 molecule containing two types of binding ligands (COOH and COO⁻) and the types of binding mechanism with TiO₂ that may occur with these binding groups. (Lee et al. 2010)

Murakoshi et al. (1995) and Falaras (1998) were the first researchers to report that the N719 photosensitzer bonded to the TiO_2 surface via ester-like bond. This was concluded from a shift in the FTIR spectrum for the $v(C=O)$ mode which was higher for the N719-adsorbed TiO_2 than for that of the pure compound. Finnie et al. (1998) contradicted previous reports (Murakoshi et al., 1995; Falaras, 1998) on the ester-like type of bonding and reported that the N3 molecule (same as N719 but has all acid groups and no carboxylates) complexes to TiO_2 in a bidentate chelating or bridging type of mode. In this case, Finnie et al. (1998) used the splitting of carboxylate stretching bands ($\Delta v = v_{asym}(COO^-) - v_{sym}(COO^-)$) to distinguish possible modes of coordination of carboxylate ion to the TiO_2. Shklover et al. (1998) made further suggestions on the anchoring modes of the N3 sensitizer on the TiO_2 based on computational studies where they proposed two thermodynamically favorable models where the N3 molecule was attached via two of its four carboxylic groups coming from two different bipyridine. More specifically, two carboxylic acid groups of N3 were thought to either bind to two adjacent rows of titanium ions through bidentate chelating coordination or interact with surface hydroxyl groups through hydrogen bonds (no supporting experimental evidence was ever presented).

Nazeeruddin et al. (2003) supported Shklover et al. (1998) findings based on their ATR-FTIR spectra where three types of Ru complexes (namely, N3, N712 and N719) were investigated to determine how many carboxylic groups participate in coordination and which groups (COOH or COO^-) are involved in the binding mechanism; the use of all four ligand groups for biding was not feasible from a steric hindrance perspective and therefore two of their four carboxylic groups were proposed to be used in binding. Based on the presence of carboxylate vibrations in the IR spectra of the N719-TiO_2 surface, they further indicated that two carboxylic groups *trans* to the NCS group involved in the binding mechanism. The possibility of unidentate coordination was ruled out, because the IR data of the N712-TiO_2 did not a show the carbonyl band at ~1700 cm^{-1}, confirming the absence of ester type of bond between the N712 (no COOH and 4 COO^-) and TiO_2. Leon et al. (2006) then investigated the N719-TiO_2 system using SERRS at 514nm, SERS at 632 nm, normal FTIR and ATR-FTIR spectroscopy's. Based on the Δv frequency change of the COO^- groups (Deacon and Philip, 1980; Srinivas et al., 2009) they concluded that the N719-TiO_2 system exhibited carboxylate groups in the bridging or bidentate chelating modes. When the 514 nm wavelength was used (this energy corresponding to the main absorption band observed in the UV-Vis spectrum "resonance"), the Raman spectra showed the presence of the $v(C=O)$ vibration which should not be observed if bidentate chelating occurs and all the carbonyl groups are displaced upon binding. The observation of this band was warranted due to the presence of adsorbed and non-adsorbed molecules which all give a Raman signal at resonance energies (~ 514nm in the case of the pure N719). The Raman spectra at 632nm (SERS but non -resonance) showed that no $v(C=O)$ vibration was present, the region of the $v(C-O)$ mode was strongly altered, and that the $v_{sym}(COO^-)$ vibration appeared; these observations were taken to be a result of the bidentate bonding that occurs upon in adsorption of the N719 which removes all the C=O bands, and only COO- vibrations remain. Recently, Hirose et al. (2008) hypothesized that the N719 dye adsorption on the TiO_2 surface was facilitated by the presence of surface OH sites that leads to the formation of bidentate chelating linkage with one of its four carboxylate groups, but no explanation was offered to why only one ligand group was chosen versus two as in the many previous studies (Finnie et al., 1998; Nazeeruddin et al., 2003; Leon et al., 2006; Kay & Gratzel, 2002; Shklover et al., 1998; Hagfelt 2010)

On the basis of the literature works reviewed above, and our own interest in hydroxyl-rich TiO_2 as photoanodes in DSSC applications, a greater understanding of the role of the N719 dye's anchoring (COO- and COOH) groups and TiO_2's surface groups (Ti-O, Ti-OH, Ti-OH_2) was needed. To this end, a series of studies was conducted on the interaction of two types of nano-crystalline anatase substrates (commercial one and our own synthetic variety) with the N719 photosensitzer and published via the use of several techniques with vibrational spectroscopy being particularly important as will be shown below (Demopoulos et al., 2009; Lee et al., 2010a, 2011a, 2011b).

4.2 Discussions and results

As can be seen from the brief literature review, the binding mechanism of these photosensitzer molecules with anatase (TiO_2) nano-crystalline substrates have been intensively investigated by numerous research groups {including those that pioneered this field (Finnie et al., 1998; Gratzel, 2001; Nazeeruddin et al., 2003; Leon et al., 2006; De Angelis et al., 2007a, 2007b; Hagfelt 2010)} around the world as a result of the fact that the binding of the dye to the TiO_2 substrate plays a key role in the efficiency of DSSC. For example, bidentate type of bonding is preferred over monodentate biding as a result of the fact it simply provides two electron bridges via the use of the covalent bonds.

Compound	$v_{asym}(COO^-)$ (cm^{-1})	$v_{sym}(COO^-)$ (cm^{-1})	Δv (cm^{-1})	Binding mode	Ref.
N719	1618	1376	242	Ester-like unidentate	Falaras et al. 1998
N719-TiO$_2$	1617	1382	235		
N719	1615	1371	251	Bridging/ Bidentate chelating	Finnie et al. 1998
N719-TiO$_2$	N/A	N/A	N/A		
N719	1608	1365	243	Bridging	Nazeeruddin et al. 2003
N719-TiO$_2$	1602	1373	227		
N719	1606	1354	252	Bridging/ Bidentate chelating	Leon et al. 2006
N719-TiO$_2$	1602	1375	227		
N719	N/A	N/A	N/A	Bidentate chelating	Hirose et al. 2008
N719-TiO$_2$	1626	1352	274		
N719	1603	1377	235	Bidentate Bridging	Lee et al. 2010
N719-TiO$_2$	1607	1377	230		

Table 4. The Δv (v $_{COO-asym}$ − v $_{COO-sym}$) values obtained for several studies on the same system and the binding mechanism suggested based on this parameter. (Deacon and Philip, 1980; Srinivas et al., 2009)

From the above review on the previous works, it became evident in almost all cases, the splitting of carboxylate stretching bands ($\Delta v = v_{asym}(COO^-) - v_{sym}(COO^-)$) was used to distinguish possible modes of coordination of the *carboxylate* ion to the TiO_2. This method was developed by Deacon and Phillip (1980) for known crystallographic carboxylate structures as a result of the low symmetry of RCO_2^- which makes the different types of carboxylate coordination indistinguishable on the basis of the number of infrared or Raman active vibration, and as a result this Δv parameter was invented. This is unlike higher symmetry molecules such as carbonates ($\sim D_{3h}$), sulfates or arsenates ($\sim T_d$) where a clear removal of degenerate modes can be observed upon the change of symmetry that occurs upon bonding of the dye molecule on the substrate. However upon a closer look to the range of Δv values for the exact same system in our studies and that of the numerous studies along the years (Table 4), it was realized that there was quite a variation in the range of values and as such this could not be taken as the only evidence of the binding mode in this system. Moreover, it was noted from the literature (Taratula et al., 2006) for other types of molecules with similar binding ligands (COOH, COO-), that upon adsorption of the ligands to the TiO_2 substrate not only was all the carbonyl band removed ($\sim$ 1700cm-1) but also the clear formation of the carboxylate bands were also observed (1300, 1600 cm-1). To verify such types of results the use of stearic acid was also investigated in parallel (Lee et al., 2010).

Most previous works over the last decades on this system have largely focused only on the dye IR and Raman vibrational structure (1000-2600 cm-1); in our investigations (Demopoulos et al., 2009; Lee et al., 2010, 2011a, 2011b), the use of the higher wavenumber region in the pure photosensitzer dye, pure nano-crystalline anatase substrates and after sensitization was investigated. As will be shown below, the analysis of the higher wavenumber region contributed greatly to the results gathered in this study.

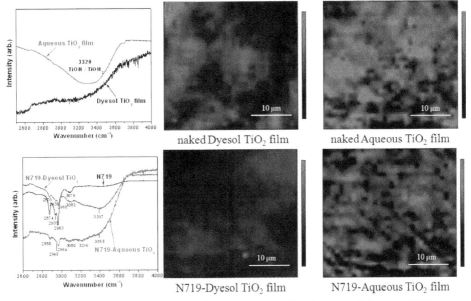

naked Dyesol TiO_2 film   naked Aqueous TiO_2 film

N719-Dyesol TiO_2 film   N719-Aqueous TiO_2 film

Fig. 12. ATR-IR spectra (Hydroxyl region) and corresponding images of the pure TiO_2 substrates (Dyesol and Aqueous) and that of the adsorbed N719-TiO_2 system. (Lee et al. 2010, 2011b)

One of the main aspects of our work (Demopoulos et al., 2009; Lee et al., 2010, 2011a, 2011b) was to compare how two similar nano-crystalline anatase TiO_2 substrates (a commercial_Dyesol and our own synthetic Aqueous product) behaved in terms of their bonding properties (via vibrational spectroscopy), distributions (via vibrational imaging) and later from an electronic interaction perspective (Lee et al., 2011b) not only from the photosensitzer (N719_Dye) perspective but also from the substrate point of view. From

Figure 12 above, it became apparent to us that the synthetic product (Demopoulos et al., 2009) produced under different conditions then the commercial TiO₂ substrate (Dyesol) contained a large number of surface OH/H₂O groups in comparison to the commercial product. This higher concentration of surface OH/H₂O groups was not only observed spectrocopically but was also confirmed visually via the use of ATR-IR imaging (Figure 12), where it was observed that indeed our synthetic aqueous nano-crystalline anatase substrate (Demopoulos et al., 2009) was indeed richer in surface OH/H₂O groups in comparison to the commercial products. As mentioned above (Hirose et al., 2008), the presence of certain OH groups has been shown to be beneficial to the binding of the dye molecule ligands and therefore increase the efficiency of the DSSC; in our case the presence of not only OH but also H₂O groups was observed spectrocopically via vibrational spectroscopy, XPS (Lee et al., 2011b) and Thermo Gravimetric Analysis. More interestingly was monitoring how these surface groups (OH/H₂O) behave after the sensitization of the N719 on the two nano-crystalline anatase substrates. From the ATR-IR spectra (Figure 12), we can observe that not all the OH group vibrations have disappeared from the spectra. This meant that not all the surface groups on the substrates are chemically active to be displaced and bind to the photosensitzer molecule, imaging of the distribution of these surface groups (Figure 12) before and after further confirmed the spectroscopic results. Moreover, we again observed that after (and before) the dye adsorption to our anatase substrate, more surface groups (OH/H₂O) were observed in the synthetic product (Demopoulos et al., 2009) relative to the commercial products, and therefore some of these non bonded surface groups must participate in the binding mechanism of the N719 molecule. These types of analysis and results (Lee et al., 2010, 2011a) had never been conducted or investigated in the previous works over the last decades; as can be observed these results simply arose from taking advantage of the entire IR region (in our case mid-IR) that one analyses which offers various information of several molecular groups in several regions of the spectra and as a result may contain important information to analyze.

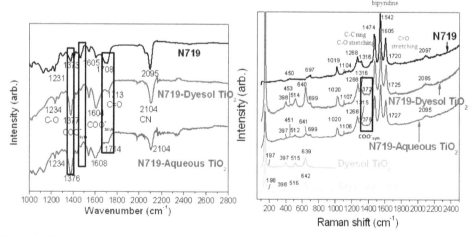

Fig. 13. ATR-IR (left) and Raman (right) spectra of the pure N719 and adsorbed onto both TiO₂ substrates [Dyesol and Aqueous]. (Lee et al. 2010)

Thus from the above results, we gathered that the Δv parameter had a large range of variation from study to study thus another type of mechanism must have been present in the binding mechanism then those previously proposed in literature studies. Moreover, we observed that the presence of surface groups remained after the adsorbtion indicating that not all these surface sites on the anatase substrates are displaced once the adsorbtion occurs. Therefore, we also decided to investigate the dye vibrational region to observe the changes that occur to the vibrational structure of the photosensitizer molecule before and after dye adsorbtion. The first thing that may be observed from the Raman and ATR-IR spectra (Figure 13) is that when we compare the pure dye (N719) spectra with that of the adsorbed states in either substrate (aqueous and Dyesol), little to no difference is observed from one state to the other unlike in other cases for molecules with similar binding ligands (Taratula et al., 2006).

More specifically in the case of the Raman spectra, we observed a carboxylate stretches which may be indicative of binding of some of the ligand molecules (COO- and/or COOH) as stated in previous works but also could simply be from the carboxylate groups not bound to the anatase surface (Figure 13). It should be noted as shown in Figure 11 above, the N719 Molecule has 4 biding ligands from which only two are reported to bind to the anatase substrate, depending whose article is read either both carboxylates or carboxylic acids are thought to be the main groups participating in the binding from an energetic point of view. Either way, two of these are carboxylates and the other two are carboxylic acids. More importantly, the Raman results showed us that there was still $C=O$ stretches observed in the spectra after the dyes adsorption on both substrates, this meant that not all the carboxylic acids groups bonded after the adsorption of the dye molecule. This in fact is an interesting fact as previous studies (Leon et al., 2006) have mentioned that the lack of this band meant that the acid groups were all bound to the substrates and stated that if observed at resonance energies (in this case 514 nm) that this is a result of getting the signal from all molecules deposited (bonded and non-bonded). To confirm these findings, we decided to probe this same system at non-resonance energies (632 nm laser probe) but at the same time avoiding emission excitation energies (~785 nm). From these measurements, we observed that indeed at non resonance energies, the presence of these $C=O$ bands was still present and as such this meant that indeed not all the acid groups were displaced upon binding of the dye molecule on the substrate (commercial and synthetic). Moreover, not only were these $C=O$ bands presented after adsorption but also were found to be shifted to higher energy relative to the pure N719 dye indicating that these groups likely were under the influence of some type of interactions (such as H-bonding with surface Ti-OH/H_2O groups as observed above) upon binding onto the anatase substrates. This latter trend (shift in energy of non bonded groups) was observed was also observed for the NCS groups of the adsorbed states relative to the pure states. In our work (Lee et al., 2010, 2011a), as mentioned above, not only were we interested in the bonding mechanism of this system but also we wanted to investigate how two otherwise similar nano-crystalline anatase substrates behaved in terms of their binding distributions throughout the photoanode. From the Raman spectral information gathered above, the use of Confocal Raman Imaging (Figure 14) was used to tune the energies of the molecular groups of interest and show the spatial distributions of these groups on these substrate before and after dye adsorption, but most importantly to observe how the bonded and non bonded groups were distributed throughout the two anatase (commercial and synthetic) substrates. For example, we decided to observe the bypiridine stretches (which definitely do not participate in bonding) to observed how the overall dye was distributed in the substrates, the COO- was used to observe where the dye was covalently bonded to the substrate and the C=O band to image how the non bonded groups from the dye molecules were distributed throughout the substrates (dark regions represented where they were bonded and bright regions where unbounded groups were observed). In all cases regardless of what group of interest was imaged, it was observed that the distribution of the dye (and therefore its molecular groups) on our synthetic substrate was unlike that of the commercial Dyesol. For our synthetic substrate, localization of the dye molecules and its bonded and non-bonded groups occurred as hot spot regions throughout the substrate, while in the case of the commercial substrate, the dye and its groups are distributed fairly evenly throughout the substrate. It should be noted to the readers that both the commercial and our synthetic anatase substrates exhibited similar roughness factors in the nanometer scale and as such these changes in distribution observed in the Raman images could not be formed as a result of their surface topography but rather from a chemical property of the materials. Thus in the case of our synthetic substrate, the localization of hot spot for the dye molecules bonding and non bonding groups on the substrate may result from the different surface energetic that occur as a result of the localization of substrate surface groups (Ti-OH/Ti-OH$_2$) which is greater in our synthetic substrate (Demopoulos et al., 2009) then the commercial Dyesol one.

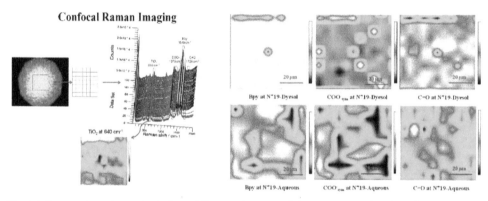

Fig. 14. Raman imaging principle and Raman images of the certain molecules of interest for the N719 molecule adsorbed onto both TiO$_2$ substrates (Dyesol_comercial and Aqueous_our synthetic). (Lee et al. 2010, 2011b)

The ATR-IR data for the dye region before and after the sensitization, similar to the Raman data showed that again from comparison of the pure dye to the adsorbed state, there was little to no change (aside from a small change in intensity) to the COO- and C=O bands (Figure 13). As mentioned above, in the case of other molecules such as the bulky m-Py-EPE-Ipa (Taratula et al., 2006) or the small stearic acid, upon binding of these molecules with the TiO$_2$ substrates, the vibrational structure of the pure molecules changes completely before and after the adsorption (i.e. the C=O band is removed and only COO- bands remain). Not only were these non bonded groups present but also a shift in energy was observed for both C=O and NCS groups, in agreement with our Raman data. Therefore, the presence of these C=O groups indicated again that not all the groups were used up in upon the adsorption of the dye and their shift in energy was a result of their interaction with the non bonded surface Ti-OH/H$_2$O groups on the surface. A similar situation was observed for the COO$^-$ groups, were the Δv parameter splitting of these before and after sensitization were indicative of bidentate bridging complexes, in agreement with previous studies but the lack of change to their vibrational structure and their energy shifts of both modes (COO$^-$sym and assym) was indicative of them interacting with the non bonded surface Ti-OH/Ti-H$_2$O groups.

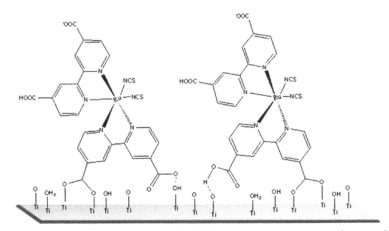

Fig. 15. New Binding mechanism for the N719-TiO system based on our vibrational spectroscopic and imaging data. (Lee et al., 2010)

Therefore, based on the vibrational spectroscopic and imaging data gathered we proposed a new modified structure for the binding mechanism of N719 dye on TiO$_2$ which has now become the standard accepted model by all the leading experts/pioneers in the field (see citations of Lee et al., 2010). Although Bazzan et al. 2011 recently published an article that stated that "there is as yet no commonly accepted understanding of the biding mechanism", in their published work, they used our proposed concepts and binding mechanism to explain the binding mechanism in their work. In our mechanism (Figure 15), the binding mechanism of the N719 dye onto TiO$_2$ was proposed to occur via two neighboring carboxylic acid/carboxylate groups linked on the same bipyridine ligand via a combination of bidentate-bridging and H-bonding, the latter part was something that had never been considered previously in the extensive research on this system. In our new binding mechanism, the formed surface covalent complex is sterically stabilized via the development of H-bonding between the TiOH/Ti-OH$_2$ (Ti-O) and COO$^-$(COOH) groups. Moreover, the involvement of only one carboxylic group in the adsorption of N719 via covalent bidentate bridging was adopted here [as in Hirose et al., (2008)] as a result of very recent work by Srinivas et al. (2009) involving the adsorption of sensitizers on TiO$_2$ carrying malonic and cyanoacrylic acid biding groups, who found that the monocarboxylic acid group showed stronger binding affinities as well as slightly higher IPCE and efficiencies than dicarboxylic acid groups, and therefore the requirement for two ligand groups to be covalently bonded versus only one was not necessary to produced higher IPCE and efficiencies.

5. Conclusions

In the first case, the Ca (II)-Fe(III)-AsO$_4$ phases (yukonite and arseniosiderite), the use of vibrational spectroscopy was employed as a finger printing tool, to correct molecular formulae via detection of certain molecular groups of interest (HAsO$_4$, H$_2$O, OH) but most importantly, it gave us an understanding to how these two distinct phases were divergent to each other at the molecular scale and why their crystallinity (long range order) differed. This was something that was previously speculated but no experimental evidence was ever really presented to explain this phenomenon. X-ray Absorption based techniques (XANES and EXAFS) of these phases (yukonite and arseniosiderite) failed to recognize this difference as a result of the fact that it could only probe three types of shells (As-O, As-Fe, As-Ca) from the core excited atom of interest. This limitation arose from the fact that these were the same at the local structural level for these two phases, based on our and previous works. Therefore the use of vibrational analysis and ATR-IR spectroscopy enabled us to reveal that it was the nature of the H-bonding molecular structure in these two phases (yukonite and arseniosiderite) which caused the difference in the physical order (crystallinity) expressed.

For the Fe(III)-AsO$_4$-SO$_4$ system, as a result of the fact that much work had been focused only on XRD based analysis, a clear understanding of the molecular structure and how this was important to the issues at hand was not yet developed. More importantly, the use of the proper detection mode (ATR vs. Transmission) for the type of material analysis at hand allowed our work to be able to extract the maximum amount of vibrational structure to be analyzed. This, along with other structural information (such as crystallographic data) was especially important when it came down to identifying possible groups (HAsO$_4$ vs. AsO$_4$) that could occur in the phase of interest which allowed us to correct previous formulae. Once the proper method of analysis was employed, the positive identification of these phases could be properly established (along with the corresponding elemental and structural data) and the identification of new phases by previous studies was eliminated.

Moreover, the use of vibrational spectroscopy gave us insight into the changes in the molecular solid-solution symmetry that occurs in the structure of a crystallographically unknown phase (BFAS), something that would be useful later when the crystal structure of this phase was solved via high resolution synchrotron diffraction data. The use of vibrational spectroscopy of industrially crystalline and non crystalline products gave us the advantage of tuning into the molecules of interest (AsO$_4$) and also offered the much lower detection limits then complementary XRD methods (4-5 w.t%) needed to analyze these samples. A nice correlation between the arsenate vibrational structure of the industrial sample and that of our synthetically produced phases was achieved but it should be noted that this information is only qualitative and 100% identification of the nature of the arsenic species in these samples is not possible using vibrational spectroscopy, XRD, or XAS based methods. This latter result arises from the fact that in the case of industrial samples where more than one, two, three arsenate phase maybe present {some which may have the same or more than one type of local arsenic coordination to the 3rd shell, e.g. yukonite-arseniosiderite}, the positive contribution to the arsenate signal from all the phases (in various coordination's and symmetries) becomes nearly impossible to qualitatively or quantitatively (via fitting or modelling) identify as a result of the fact that the contribution from each phase to the overall signal and features is almost impossible to replicate. This is especially true at low concentrations and in the presence of a multicomponent matrix which can also have additional effects to the signal observed. Moreover, limitations of vibrational spectroscopy were observed in cases where phases such as hematite and elemental sulfur (which have large Raman cross sections) were abundant in industrial products. There Raman spectroscopy was only able to detect one of these phases in spite of the fact that the other was well in abundance of the Raman detection limit. This further shows the reader, the complexity that a multicomponent sample has on the vibrational structure expressed in the current spectroscopic techniques.

Finally, for the N719-TiO_2 biding mechanism, discrepancies in the coordination information inferred based on the Δv method for the same system led us to further evaluate the detailed vibrational structure of the system. In this case, the combination of the spectroscopic information coupled with the aid of the imaging data allowed us to determine how the molecular groups of interest on the TiO2 substrates and N719 molecule behaved during the binding event, but also how these were distributed among the two TiO_2 substrates investigated. Using the information of previous works along with the new gathered data from our studies, the evolution of a new binding mechanism which involved a bridged covalent bonded-H-bonding type of mechanism was developed for this system. The main draw back or limitation of the use of vibrational spectroscopy in this study was the fact that the imaging of the substrates was limited to micro-scale analysis as a result of the laser beam used; however, it is well known for these systems that the internal surfaces are uniformly covered by a monolayer of dye, and a such the use of nano-imaging techniques such as AFM-Raman analysis may give more insight into how these groups are distributed among the various substrates at the nano-level but also probe the changes of the molecular groups of interest at the nano-scale in which bulk like effects may be avoided.

In this chapter, three distinct case studies were presented to the reader where the use and application of vibrational spectroscopy in correlation with other analytical data was employed to give new insights into the various problems at hand. Most importantly, the limitations of the techniques (vibrational, XRD, XAS) employed in previous studies and in our work was presented to show how these restrictions may be overcome by the combination of various techniques, detection methods and a good knowledge of the systems theory at hand. However, the reader is forewarned to be cautious when using vibrational spectroscopy (XAS or XRD) to identify multiple phases in a complex matrix; often specialist in these techniques or research facilities (e.g. synchrotron facilities) offer that each one of these can offer a solution to identify all of the main phase of interest found in a complex sample but this is often not true and is beyond the technological (and theoretical) capabilities of these techniques and facilities. A perfect way to test such a concept is to test a known multiphase (> 4) poorly crystalline industrial or natural sample and see if indeed any of these techniques (vibrational, X-ray Absorption Spectroscopy or X-ray Diffraction) identify all the phases in the sample, naturally they will not for now.

6. Acknowledgments

The authors would like to thank several instrumental people
from various universities and research facilities who were co-authors of these published works or gave their time, funding and efforts to these studies: Prof. George P.Demopoulos and Dr. Jeff N. Cutler Dr. Samir Elouatik, Dr. Hassane Assoudi, Dr. Leventer Becze, Dr. Gerry Ventruti, Petr Fiurasek, Xu Dong Li, Mert Celikin,Tom Regier, Dr. R. I. R. Blyth, Dr. Yonfeng Hu. We also like to thank Joan Kaylor from the Redpath Museum at McGill University for the donation of the Romanech Arseniosiderite sample and Dr. Richard Herd, Curator of the National Collections of Canada for the donation of the original natural Tagish

Lake Yukonite sample. Research described in this book chapter for the X-ray Absorption Analysis was performed at the Canadian Light Source which is acknowledged.

7. References

Bastians, S.; Crump, G.; Griffith, W. P.; Withnall, R. (2004) Raspite and studtite: Raman spectra of two unique minerals. *J. Raman Spectrosc.* Vol. 35, No. 8-9, (January 2004), pp. 726-731, ISSN: 0377-0486.

Bazzan, G.; Deneault, J. R.; Kang, T.K.; Taylor, B.E.; Durstock, M.F. (2011) Nanoparticle/Dye Interface Optimization in Dye-Sensitized Solar Cells. *Adv. Funct. Mater.* Vol. 21, No. 17, (Septemeber 2011), pp. 3268-3274, ISSN: 1616-3028.

Becze, L.; Gomez, M. A.; Petkov, V.; Cutler, J. N.; Demopoulos, G. P. (2010) The Potential Arsenic Role of Ca-Fe(III)-AsO4 Compounds in Lime Neutralized Co-Precipitation Tailings, *Proceedings of Uranium 2010 and 40th Annual Hydrometallurgy Meeting, Uranium Processing*, Vol. II, pp. 327-336, Saskatoon, SK, Canada, August 2010.

Becze, L.; Demopoulos, G. P. (2007) Synthesis and Stability Evaluation of Ca-Fe(III)-AsO4 Phases, *Extraction and Processing Proceedings*, B. Davis and M. Free, Eds., TMS, pp. 11-17, ISBN: 978-0-470-93163-9.

Becze, L.; Gomez, M. A.; Le Berre, J. F.; Pierre, B.; Demopoulos, G. P. (2009) Formation of Massive Gunningite-Jarosite Scale in an Industrial Zinc Pressure Leach Autoclave: a Characterization Study. *Can. Metall. Q.* Vol. 48, No. 2, (June 2009), pp. 99-108, ISSN: 0008-4433.

Berezowsky, R.; Xue, T.; Collins, M.; Makwana, M.; Barton-Jones, I.; Southgate, M.; Maclean, J. (1999) Pressure leaching las cruces copper ore, *JOM*, Vol. 51, No. 12, (December 1999), pp. 36-40, ISSN: 1047-4838.

Bruce, R.; Mayhew, K.; Mean, R.; Kadereit, H.; Nagy A.; Wagner, O. (2011) Unlocking Value in Copper Arsenic Sulphide Resources with Copper-Arsenic CESL Technology, *Hydro-copper Conference Proceeding.* Viña del Mar, Chile, July 2011

Burns, P.C.; Hughes, K. A. (2003) Studtite, [(UO₂)(O₂)(H₂O)₂](H₂O)₂: The first structure of a peroxide mineral. *Am. Mineral.* Vol. 88, No. 7, (July 2003), pp. 1165-1168, ISSN: 1945-3027.

Charbonneau, C.; Lee, K. E.; Shan, G. B.; Gomez, M. A.; Gauvin, R.; Demopoulos, G. P. (2010) Preparation and DSSC performance of mesoporous film photoanodes based on hydroxyl-rich anatase nanocrystallites. *Electrochem. Solid-State Lett.*, Vol. 13, No. 8, (May 2010), pp. H257-H260, ISSN: 1944-8775.

Coates, J. (1998) Vibrational Spectroscopy: Instrumentation for Infrared and Raman Spectroscopy. *Appl. Spectrosc. Rev.* Vol. 33, No. 4, (August 2006), pp. 267-425, ISSN: 1520-569X.

Dana, J. D.; Dana, E. S.; Palache, C.; Berman, H.; Frondel, C. (1976) *The System of Mineralogy of Dana*, Seventh ed., Vol. 2, John Wiley and Sons, Inc., ISBN: 0471193100, New York, pp. 955-956.

Deacon, G. B.; Phillips, R. J. (1980) Relationships between the carbon-oxygen stretching frequencies of carboxylato complexes and the type of carboxylate coordination. *Coord. Chem. Rev.* Vol. 33, No. 3, (October 1980), pp. 227–250, ISSN: 0010-8545.

De Angelis, F.; Fantacci, S.; Selloni, A.; Nazeeruddin, Md. K.; Gratzel, M. (2007a) Time-Dependent Density Functional Theory Investigations on the Excited States of

Ru(II)-Dye-Sensitized TiO$_2$ Nanoparticles: The Role of Sensitizer Protonation. *J. Am. Chem. Soc.* Vol. 129, No. 46, (October 2007), pp. 14156–14157, ISSN: 0002-7863.

De Angelis, F.; Fantacci, S.; Selloni, A.; Gratzel., M.; Nazeeruddine, Md. K. (2007b) Influence of the Sensitizer Adsorption Mode on the Open-Circuit Potential of Dye-Sensitized Solar Cells. *Nano Lett.* Vol. 7, No. 10, (September 2007), pp. 3189–3195. ISSN: 1530-6992.

Defreyne, J.; Cabra, T. (2009) Early copper production results from Vale's hydrometallurgical CESL refinery, *ALTA Conference Proceedings*, Perth, Australia, May 2009.

de Groot, F. M. F.; Fuggle, J. C.; Thole, B. T.; Sawatzky, G. A. (1990) 2p x-ray absorption of 2d transition-metal compounds: An atomic multiplet description including the crystal field. *Phys. Rev. B: Condens. Matter*, Vol. 42, No. 9, (September 1990), pp. 5459-5468, ISSN: 1550-235X.

de Groot F. M. F. (2005) Multiplet effects in X-ray spectroscopy. *Coord. Chem. Rev.* Vol. 249, No. 1-2, (January 2005), pp. 31-63, ISSN: 0010-8545.

Demopoulos, G. P.; Charbonneau, C.; Lee, K. E.; Shan, G. B.; Gomez, M. A.; Gauvin, R. (2009) Synthesis of Hydroxyl-Rich Anatase Nanocrystallites, Their Characterization and Performance as Photoanode in Dye-Sensitized Solar Cells. *ECS Trans.* Vol. 21, No. 1, (July 2009), pp. 23–34, ISSN: 1938-6737.

Drahota, P.; Rohovec, J.; Filippi, M.; Mihaljevic, M.; Rychlovsky, P.; Cerveny, V.; Pertold, Z. (2009) Mineralogical and geochemical controls of arsenic speciation and mobility under different redox conditions in soli, sediment and water at the Mokrsko-West gold deposit, Czech Republic. *Sci. Total. Environ.* Vol. 407, No. 10, (May 2009), pp. 3372- 3384, ISSN: 0048-9697.

Drahota, P.; Filippi, M. (2009) Secondary arsenic minerals in the environment: A review. *Environ. Int.* Vol. 35, No. 8, (November 2009), pp. 1243-1255, ISSN: 0160-4120.

Dunn, P. J. (1982) New data for pitticite and a second occurrence of yukonite at Sterling Hill, New Jersey. *Mineral. Mag.* Vol. 46, pp. 261-264, ISSN: 0026-461X.

Dutrizac, J. E.; Jambor, J. L. (2007) Characterization of the iron arsenate-sulfate compounds precipitated at elevated temperatures. *Hydrometallurgy.* Vol. 86, No. 3-4, (May 2007), pp. 147-163, ISSN: 0304-386X.

Dymov, I.; Ferron, C. J.; Phillips, W. (2004) Pilot Plant Evaluation of a Hybrid Biological Leaching-Pressure Oxidation Process for Auriferous Arsenopyrite/Pyrite Feedstocks. *In Pressure Hydrometallurgy* (eds. Collins M. J. and Papangelakis V.G.). CIM, pp. 735-750, Banff, Canada, October 2004.

Falaras, P. (1998) Synergetic effect of carboxylic acid functional groups and fractal surface characteristics for efficient dye sensitization of titanium oxide. *Sol. Energy Mater. Sol. Cells.* Vol. 53, No. 1-2, (May 1998), pp. 163–175, ISSN: 0927-0248.

Fernández-Martínez, A.; Cuello, G. J.; Johnson, M. R.; Bardelli, F.; Román-Ross, G.; Charlet, L.; Turrillas, X. (2008) Arsenate Incorporation in Gypsum Probed by Neutron, X-ray Scattering and Density Functional Theory Modeling. *J. Phys. Chem. A*, Vol. 112, No. 23, (May 2008), pp. 5159-5166, ISSN: 1089-5639.

Filippi, M.; Golias, V.; Pertold, Z. (2004) Arsenic in contaminated soils and anthropogenic deposits at the Mokrsko, Roudny, and Kasperske Hory gold deposits. *Environ. Geol.* Vol. 45, No. 5, (March 2004), pp. 716-730, ISSN: 1432-0495.

Filippi, M.; Dousova, B.; Machovic, V. (2007) Mineralogical speciation of arsenic in soils above the Mokrsko-west gold deposits, Czech Republic. *Geoderma*. Vol. 139, No. 1-2, (April 2007), pp. 154-170, ISSN: 0016-7061.

Fillipou, D.; Demopoulos, G. P. (1997) Arsenic Immobilization by Controlled Scorodite Precipitation. *JOM*. Vol. 49, No. 12, (December 1997), pp. 52-55, ISSN: 1047-4838.

Finnie, K. S.; Bartlett, J. R.; Woolfrey, J. L. (1998) Vibrational Spectroscopic Study of the Coordination of (2,2'-Bipyridyl-4,4'-dicarboxylic acid)ruthenium(II) Complexes to the Surface of Nanocrystalline Titania. *Langmuir*. Vol. 14, No. 10, (February 1998), pp. 2744-2849, ISSN: 0743-7463.

Fleet, M. E.; Liu, X. (2009) Calcium $L_{2,3}$-edge XANES of carbonates, carbonate apatite and oldhamite (CaS). *Am. Mineral*. Vol. 94, No. 8-9, (August-September 2009), pp. 1235-1241, ISSN: 1945-3027.

Foshac, W. F. (1937) Carminite and associated minerals from Mapimi, Mexico. *Am. Mineral*. Vol. 22, No. 5, (May 1937), pp. 479-484, ISSN: 1945-3027.

Frost, R.; Weier, M.; Martens, W. (2006) Raman microscopy of synthetic goudeyite ($YCu_6(AsO_4)_2(OH)_6 3H_2O$). *Spectrochim. Acta A*. Vol. 63, No. 3, (March 2006), pp. 685–689, ISSN: 1386-1425.

Fujita, T.; Taguchi, R.; Abumiya, M.; Matsumoto, M.; Shibata, E.; Nakamura, T. (2008) Novel atmospheric scorodite synthesis by oxidation of ferrous sulfate solution. Part I *Hydrometallurgy*, Vol. 90, No. 2-4, (February 2008), pp. 92-102, ISSN: 0304-386X.

Garavelli, A.; Pinto, D.; Vurro, F.; Mellini, M.; Viti, C.; Balic-Zuni, T.; Della Ventura, G. (2009) Yukonite from the Grotta della Monaca Cave, Sant'Agata Di Esaro, Italy: Characterization and comparison with Cotype material from the Daulton ine, Yukon, Canada. *Can. Mineral*. Vol. 47, No. 1, (February 2009), pp. 39-51, ISSN: 0008-4476.

Gomez, M. A.; Becze, L.; Blyth, R. I. R.; Cutler, J. N.; Demopoulos, G. P. (2010a) Molecular and structural investigation of yukonite (synthetic & natural) and its relation to arseniosiderite. *Geochim. Cosmochim. Acta*. Vol 74, No. 20, (October 2010), pp. 5835-5851, ISSN: 0016-7037.

Gomez, M. A.; Assaaoudi, H.; Becze, L.; Cutler, J. N.; Demopoulos, G. P. (2010b) Vibrational spectroscopy study of hydrothermally produced scorodite ($FeAsO_4 \cdot 2H_2O$), ferric arsenate sub-hydrate (FAsH; $FeAsO_4 0.75 H_2O$) and basic ferric arsenate sulphate (BFAS; $Fe[(AsO_4)_{1-x}(SO_4)_x(OH)_x] \cdot wH_2O$) *J. Raman Spectrosc*. Vol. 41, No.2, (February 2010), pp. 212-221, ISSN: 0377-0486.

Gomez, M. A.; Becze, L.; Cutler, J. N.; Demopoulos, G. P. (2011a) On the hydrothermal reaction chemistry and characterization of ferric arsenate phases precipitated from $Fe_2(SO_4)_3$-As_2O_5-H_2SO_4 solutions. *Hydrometallurgy*. Vol. 107, No. 3-4, (May 2011), pp. 74-90, ISSN: 0304-386X.

Gomez, M. A.; Le Berre, J. F.; Assaaoudi, H.; Demopoulos, G. P. (2011b) Raman spectroscopic study of the hydrogen and arsenate bonding environment in isostructural synthetic arsenates of the variscite group-$M^{3+}AsO_4 \cdot 2H_2O$ (M^{3+} = Fe, Al. In and Ga)-implications for arsenic release in water. *J. Raman Spectrosc*. Vol. 42, No. 1, (January 2011), pp. 62-71, ISSN: 0377-0486.

Gomez, M. A.; Becze, L.; Celikin, M.; Demopoulos, G. P. (2011c) The effect of copper on the precipitation of scorodite ($FeAsO_4 2H_2O$) under hydrothermal conditions: Evidence for a hydrated copper containing ferric arsenate sulfate-short lived intermediate. *J. Colloid Interface Sci.* Vol. 360, No. 2, (August 2011), pp. 508-518, ISSN: 0021-9797.

Gomez, M. A.; Ventruti, G.; Assaaoudi, H.; Ceklin, M.; Putz, H; Lee, K. E.; Demopoulos, G. P. (2011d) The nature of synthetic Basic Ferric Sulfate and Basic Ferric Arsenate Sulfate: A crystal and molecular structure determination and applications of their materials properties. (*Submitted*)

Gratzel, M. (2001) Photoelectrochemical cells. *Nature*. Vol. 414, (November 2001), pp. 338-344, ISSN: 0028-0836.

Guana, X.; Wang, J.; Chusue. C. C. (2008) Removal of arsenic from water using granular ferric hydroxide: Macroscopic and microscopic studies. *J. Hazard. Mater.* Vol. 156, No. 1-3, (August 2008), pp. 178-185, ISSN: 0304-3894.

Hagfeldt, A.; Boschloo, G.; Sun, L.; Kloo, L.; Pettersson, H. (2010) Dye-sensitized Solar Cells. *Chem. Rev.* Vol. 110, No. 10, (September 2010), pp. 6595-6663, ISSN: 0009-2665.

Harris, D. C.; Bertolucci, M. D. (1989) *Symmetry and Spectroscopy: An Introduction to Vibrational and Electronic Spectroscopy.* Dove Publications, Inc., ISBN: 048666144X, New York.

Hawthorne, F. C. (1976) The Hydrogen Positions in Scorodite. *Acta Crystallogr. Sect. B: Struct. Sci.* Vol. 32, No.10, (October 1976), pp. 2891-92, ISSN: 0108-7681.

Hirose, F.; Kuribayashi, K.; Suzuki, T.; Narita, Y.; Kimura, Y.; Niwano, M. (2008) UV Treatment Effect on TiO_2 Electrodes in Dye-Sensitized Solar Cells with N719 Sensitizer Investigated by Infrared Absorption Spectroscopy. *Electrochem. Solid-State Lett.* Vol. 11, No. 12, (October 2009), pp. A109-A111, ISSN: 1944-8775.

Hollas, J. M. (2004) *Modern Spectroscopy.* (6th edn), Wiley, ISBN: 0470844167, New York.

Jakeman, R. J. B.; Kwiecien, M. J.; Reiff, W. M.; Cheetham, K.; Torardi, C. C. (1991) A new ferric orthoarsenate hydrate: structure and magnetic ordering of $FeAsO_4 \cdot 3/4H_2O$. *Inorg. Chem.* Vol. 30, No. 13, (June 1991), pp. 2806-2811, ISSN: 0020-1669.

Jambor, J. L. (1966) Abstract of paper presented at the eleventh meeting: Re-examination of yukonite. *Can. Mineral.* Vol. 8, pp. 667-668, ISSN: 0008-4476.

Jia, Y.; Demopoulos, G. P. (2005) Adsorption of Arsenate onto Ferrihydrite from Aqueous Solution: Influence of Media (Sulphate vs Nitrate), Added Gypsum, and pH Alteration. *Enviro. Sci.Technol.* Vol. 39, Vol. 24, (November 2005), 9523-9527. ISSN: 0013-936X.

Jia, Y.; Demopoulos, G. P. (2007) Coprecipitation of arsenate with iron(III) in aqueous sulfate media: effect of time, lime as base and co-ions on arsenic retention. *Water Res.* Vol. 42, No. 3, (August 2007) pp. 661-6688. ISSN: 0043-1354.

Kay, A.; Gratzel, M. (2002) Dye-Sensitized Core–Shell Nanocrystals: Improved Efficiency of Mesoporous Tin Oxide Electrodes Coated with a Thin Layer of an Insulating Oxide. *Chem. Mater.* Vol. 14, No. 7, (June 2002), pp. 2930-2935, ISSN: 0897-4756.

Koenig, G. A. (1889) VI. Neue amerikanische Mineralvorkommen.1. Mazapilit. *Z. Kristallogr.* Vol. 17, pp. 85-88, ISSN: 0044-2968.

Krause, E.; Ettel, V. A. (1989) Solubilities and stabilities of ferric arsenate compounds. *Hydrometallurgy.* Vol. 22, No. 3, (August 1989), pp. 311-337, ISSN: 0304-386X.

Kroon, J. M.; Bakker, N. J.; Smit, H. J. P.; Liska, P.; Thampi, K. R.; Wang, P.; Zakeeruddin, S. M.; Gratzel, M.; Hinsch, A.; Hore, S.; Wurfel, U.; Sastrawan, R.; Durrant, J. R.; Palomares, E.; Pettersson, H.; Gruszecki, T.; Walter, J.; Skupien, K.; Tulloc, G. E.

(2007) Nanocrystalline dye-sensitized solar cells having maximum performance. *Prog. Photovolt: Res. Appl.* Vol. 15, No. 1, (January 2007), pp. 1–18, ISSN: 1099-159X.

Langmuir, D.; Mahoney, J.; MacDonald, A.; Rowson, J. (1999) Predicting arsenic concentrations in the pore waters of buried uranium mill tailings. *Geochim. Cosmochim. Acta.* Vol. 63, No. 19-20, (October 1999), pp. 3379-3394, ISSN: 0016-7037.

Langmuir, D.; Mahoney, J.; Rowson, J. (2006) Solubility products of amorphous ferric arsenate and crystalline Scorodite ($FeAsO_4 \cdot 2H_2O$) and their application to arsenic behavior in buried mine tailings. *Geochim. Cosmochim. Acta.* Vol. 70, No. 12, (June 2006), pp. 2942–2956, ISSN: 0016-7037.

Larsen, E. S. III (1940) Overite and montgomeryite: two new minerals from Fairfield, Utah. *Am. Mineral.* Vol. 25, No. 5, (May 1940), pp. 315-326, ISSN: 1945-3027.

Lee, K. E.; Gomez, M. A.; Elouatik, S.; Demopoulos, G. P. (2010) Further understanding of the adsorption of N719 complex on Anatase TiO_2 films for DSSC applications using Vibrational spectroscopy and Confocal Raman imaging. *Langmuir*, Vol. 26, No. 12, (March 2010), pp. 9575-9583, ISSN: 0743-7463.

Lee, K. E.; Gomez, M. A.; Regier, T.; Hu, Y.; Demopoulos, G. P. (2011a) Electronic Studies of N719/TiO_2 interface for DSSC applications: A combined XAS and XPS study. *J. Phys. Chem. C.* Vol. 115, No. 13, (March 2011), pp. 5692-5707, ISSN: 1932-7447.

Lee, K. E.; Gomez, M. A.; Shan, G. B.; Demopoulos, G. P. (2011b) Vibrational Spectroscopic Imaging of N719-TiO_2 Films in the High Wavenumber Region Coupled to Electrical Impedance Spectroscopy. *J. Electrochem. Soc.*, Vol. 158, No. 7, (May 2011), pp. 1-7, ISSN: 0013-4651.

Leon, C. P.; Kador, L.; Peng, B.; Thelakkat, M. (2006) Characterization of the Adsorption of Ru-bpy Dyes on Mesoporous TiO_2 Films with UV–Vis, Raman, and FTIR Spectroscopies. *J. Phys. Chem. B.* Vol. No. 17, (October 2005), 110, pp. 8723–8730, ISSN: 1089-5647.

Libowitsky, E. (1999) Correlation of O-H stretching frequencies and OH•••O hydrogen bond lengths in minerals. *Monatsh. Chem.* Vol. 130, No. 8, (August 1999), pp. 1047-1059, ISSN: 1434-4475.

Mayhew, K.; Parhar, P.; Salomon-de-Friedberg, H. (2010) CESL Process as Applied to Enargite-Rich Copper Concentrates. *Proceedings of the 7th International Copper-Cobre Conference*, Vol. 5, pp. 1983-1998, Hamburg, Germany, June 2010.

Moore, P. A.; Ito, J. (1974) I. Jahnsite, segelerite, and robertsite, three new transition metal phosphate species. II. Redefinition of overite, an isotype of segelerite. III. Isotypy of robertsite, mitridatite, and arseniosiderite . *Am. Mineral.* Vol. 59, No. 1-2, (January – February 1974), pp. 48–59, ISSN: 1945-3027.

Moore, P. A.; Araki, T. (1977a) Mitridatite, $Ca_6(H_2O)_6[Fe^{III}_9O_6(PO_4)_9] \cdot 3H_2O$. A noteworthy octahedral sheet structure. *Inorg. Chem.* Vol. 16, No. 5, (May 1977), pp. 1096–1106, ISSN: 0020-1669.

Moore, P. A.; Araki, T. (1977b) Mitridatite: a remarkable octahedral sheet structure. *Mineral. Mag.* Vol. 41, (December 1977), pp. 527–528, ISSN: 0026-461X.

Murakoshi, K.; Kano, G.; Wada, Y.; Yanagida, S.; Miyazaki, H.; Matsumoto, M.; Murasawa, S. (1995) Importance of binding states between photosensitizing molecules and the TiO_2 surface for efficiency in a dye-sensitized solar cell. *J. Electroanal. Chem.* Vol. 396, No. 1-2, (October 1995), pp. 27-34, ISSN: 1945-7111.

Myneni, S. C. B.; Traina, S. J.; Waychunas, G. A.; Logan, T. J. (1998) Experimental and theoretical vibrational spectroscopic evaluation of arsenate coordination in aqueous solutions, solids, and at mineral-water interfaces. *Geochim. Cosmochim. Acta*. Vol. 62, No. 19-20, (October 1998), pp. 3285–3661, ISSN: 0016-7037.

Nakamoto, K. (2009) *Infrared and Raman Spectra of Inorganic and Coordination Compounds. Part A: Theory and Applications in Inorganic Chemistry* (6th edn), Wiley, ISBN: 0471163929, New York.

Nazeeruddin, Md. K.; Humpry-Baker, R.; Liska, P.; Gratzel, M. (2003) Investigation of Sensitizer Adsorption and the Influence of Protons on Current and Voltage of a Dye-Sensitized Nanocrystalline TiO_2 Solar Cell. *J. Phys. Chem. B*. Vol. 107, No. 34, (March 2003), pp. 8981–8987, ISSN: 1089-5647.

Nishikawa, O.; Okrugin, V.; Belkova, N.; Saji, I.; Shiraki, K.; Tazaki, K. (2006) Crystal symmetry and chemical composition of yukonite: TEM study of specimens collected from Nalychevskie hot springs, Kamchatka, Russia and from Venus mine, Yukon Territory, Canada. *Mineral. Mag*. Vol. 70, No. 1, (February 2006), pp. 73-81, ISSN: 0026-461X.

Paktunc, D.; Foster, A.; Laflamme, G. (2003) Speciation and characterization of arsenic in Ketza river mine tailings using X-ray absorption spectroscopy. *Environ. Sci. Techn*. Vol. 37, No., (10), (May 2003), pp. 2067-2074, ISSN: 0013-936X.

Paktunc, D.; Foster, A.; Heald, S.; Laflamme, G. (2004) Speciation and characterization of arsenic in gold ores and cyanidation tailings using X-ray absorption spectroscopy. *Geochim. Cosmochim. Acta*. Vol. 68, No. 5, (March 2004), pp. 969-983, ISSN: 0016-7037.

Peter, L. M. *(2011)* The Gratzel Cell: Where Next? *J. Phys. Chem. Lett.*, Vol. 2, No. 15, (July 2011), pp 1861–1867, ISSN: 1948-7185.

Pieczka, A.; Golobiowska, B.; Franus, W. (1998) Yukonite, a rare Ca-Fe arsenate, from Redziny (Sudetes, Poland). *Eur. J. Mineral*. Vol. 10, No. 6, (July 1998), pp. 1367-1370, ISSN: 0935-1221.

Riveros, P. A.; Dutrizac, J. E.; Spencer, P. (2001) Arsenic Disposal Practices in the Metallurgical Industry. *Can. Metall. Q.* Vol. 40, No. 4, pp. 395-420, ISSN: 0008-4433.

Ross, D. R.; Post, J. E. (1997) New data on yukonite. *Powder Diffr*. Vol. 12, pp. 113-116, ISSN: 0885-7156.

Schwendtner, K.; Kolitsch, U. (2007) *Eur. J. Mineral*. Vol. 19, No. 3, (May-June 2007), pp. 399–409, ISSN: 0935-1221.

Scrosati, B.; Garche, J. (2010) Lithium batteries: Status, prospects and future. *J. Power Sources*. Vol. 195, No. 9, (May 2010), pp. 2419-2430, ISSN: 0378-7753.

Shklover, V.; Ovchinnikov, Y. E.; Braginsky, L. S.; Zakeeruddine, S. M.; Gratzel, M. (1998) Structure of Organic/Inorganic Interface in Assembled Materials Comprising Molecular Components. Crystal Structure of the Sensitizer Bis[(4,4'-carboxy-2,2'-bipyridine)(thiocyanato)]ruthenium(II). *Chem. Mater*. Vol. 10, No. 9, (July 1998), pp. 2533–2541, ISSN: 0897-4756.

Singhania, S.; Wang, Q.; Filippou, D.; Demopoulos, G. P. (2005) Temperature and seeding effects on the precipitation of scorodite from sulfate solutions under atmospheric-pressure conditions. *Metall. Mater. Trans. B*. Vol. 36, No. 3, (June 2005), pp. 327–333, ISSN: 1543-1916.

Srinivas, K.; Yesudas, K.; Bhanuprakash, K.; Rao, V. J.; Girbabu, L. (2009) A Combined Experimental and Computational Investigation of Anthracene Based Sensitizers for DSSC: Comparison of Cyanoacrylic and Malonic Acid Electron Withdrawing Groups Binding onto the TiO_2 Anatase (101) Surface. *J. Phys. Chem. C.* Vol. 113, No. 46, (October 2009), pp. 20117–20126, ISSN: 1932-7447.

Swash, P. M.; Monhemius; A. J. (1994) Hydrothermal precipitation from aqueous solutions containing iron (III), arsenate and sulfate, In *Hydrometallurgy '94*, ISBN 978-0-412-59780-0, Cambridge, England, July 1994.

Swash, P. M. (1996) The hydrothermal precipitation of Arsenic solids in the $Ca-Fe-AsO_4-SO_4$ system at elevated temperature, *Ph.D. Dissertation*, Imperial College of Science, Technology and Medicine, University of London, London.

Taratula, O.; Rochford, J.; Piotrowiak, P.; Galoppini, E. (2006) Pyrene-Terminated Phenylenethynylene Rigid Linkers Anchored to Metal Oxide Nanoparticles. *J. Phys. Chem. B.* Vol. 110, No. 32, (July 2006), pp. 15734–15741, ISSN: 1089-5647.

Tyrrell, J. B.; Graham, R. P. D. (1913) Yukonite, a new hydrous arsenate of iron and calcium, from Tagish Lake, Yukon Territory, Canada; with a note on the Associated Symplesite. *Trans. R. Soc. Can.*, section. IV. pp. 13–18, ISSN: 0035-9122.

Ugarte, F. J. G.; Monhemius, A. J. (1992) Characterization of high temperature arsenic containing residues from hydrometallurgical processes. *Hydrometallurgy.* Vol. 30, No. 1-3, (June 1992), pp. 69–86, ISSN: 0304-386X.

Walker, S. R.; Jamieson, H. E.; Lanzirotti, A.; Andrade, C. F.; Hall, G. E. M. (2005) The speciation of arsenic in iron oxides in mine wastes from the giant gold mine, N.W.T.: application of synchrotron micro-XRD and micro-XANES at the grain scale. *Can. Mineral.* Vol. 43, No. 4, (August 2005), pp. 1205-1224, ISSN: 0008-4476.

Walker, S. R.; Parsons, M. B.; Jamieson, H. E.; Lanzirotti, A. (2009) Arsenic mineralogy of near-surface tailings and soils: Influences on arsenic mobility and bioaccessibility in the Nova Scotia gold mining districts. *Can. Mineral.* Vol. 47, No. 3, (June 2009), pp. 533-556, ISSN: 0008-4476.

Wang, W.; Lin, Z. (2010) Dye-Sensitized TiO_2 Nanotube Solar Cells with Markedly Enhanced Performance via Rational Surface Engineering. *Chem. Mater.* Vol. 22, No. 2, (December 2009), pp. 579–584, ISSN: 0897-4756.

2

Vibrational Spectroscopy of Gas Phase Functional Molecules and Their Complexes Cooled in Supersonic Beams

Takayuki Ebata, Ryoji Kusaka and Yoshiya Inokuchi
Hiroshima University,
Japan

1. Introduction

Functional molecules and supramolecules are the assemble of molecules, which are bound by noncovalent interactions, such as dispersion force, coordinate-bonding, hydrogen-bonding, etc. They exhibit special functions by forming regular high dimensional structures controlled by those weak interactions. The concept of the supramolecule was first proposed by Jean-Marie Lehn, who succeeded in synthesizing cryptand (Lehn, 1995). In the early stage, many studies have been carried out for host-guest complexes of crown ether (Gokel, 1991; Izatt et al., 1969; Pedersen, 1967; Pedersen & Frensdorff, 1972), calixarene (Atwood et al., 2002; Gutsche, 1998; Purse et al., 2005; Thallapally et al., 2005), and cyclodextrin (Brocos et al., 2010; Szejtli, 1988). These studies are extended to larger size systems built by several units, such as protein, Langmuir- Blodgett (LB) film, self assembled monolayer (SAM), and liquid crystal. In addition, more complicated molecular assemblies, rotaxane, catenane, and molecular capsules, are synthesized. Also, many functional groups have been used to applications such as sensing and basic chemical research, by covalently linking fluorescent dyes, nanoparticles, proteins, DNA, and other compounds of interest. Biomolecules may also be categorized to the functional molecules. In the biomolecules, many units form high dimensional structure through noncovalent interactions and exhibit special functions which are not possible for each unit.

Among many experimental studies on the functional molecules, vibrational spectroscopy is one of the most general methods. By examining the frequency shift and intensity change of the specific vibrations sensitive to the local interaction, we can study which part or site of the molecule bound for the complexation. In this sense, it is necessary to investigate vibrational spectra not only of the complexes but also of each unit under the isolated condition. By the comparison of the spectra at different conditions, we understand how the molecules change their initial structures or which conformer is preferred for the complexation. A problem for this study is that in most cases they have flexible structures so that the structures are affected by many factors, such as temperature, local environment, solvent molecules as well as the phases. These effects result in the homogenous and inhomogeneous broadening of the spectra, which leads to the difficulty to analyze the spectra. In this chapter, we describe the vibrational spectroscopic study of gas phase small size functional molecules cooled in the supersonic jet. The supersonic jet technique enables us to cool the internal temperature of molecules and complexes. We apply several spectroscopic methods to these molecules. Laser induced fluorescence (LIF), one-photon resonant two-photon ionization (R2PI) and ultraviolet-ultraviolet hole burning (UV-UV HB) spectroscopic methods are used to obtain the S_1-S_0 electronic spectrum and discriminate different species, such as conformers and isomers. The infrared (IR) absorption spectrum is obtained by IR-UV double resonance (IR-UV DR) spectroscopy, which was developed to obtain the IR spectra of selected species with low concentration. IR photodissociation (IRPD) spectroscopy is another version of the IR spectroscopic measurement. By comparing the IR-UV DR and IRPD spectra, we can obtain the dissociation energy of the complex.

We first review our vibrational spectroscopic study on the conformation of L-phenylalanine (L-Phe) and L-tyrosine (L-Tyr), and the structure of the hydrated complexes. We show the discrimination of different conformers by IR-UV DR spectroscopy and how the initial geometry changes when they form complexes with water molecules. We then review the study on the encapsulation complexes of dibenzo-18-crown-6-ether (DB18C6), benzo-18-crown-6-ether (B18C6), calix[4]arene (C4A). For crown ethers (CEs), we first examine the possible conformations in bare form and then investigate how the CEs change their conformation to encapsulate the guest species in their cavities. We also comment on the molecular recognition in the process of the encapsulation. For C4A, we investigate encapsulation complexes with variety of guest species to extract which noncovalent interaction is the major component to encapsulate each guest species.

For the determination of the conformation as well as the complex structures, high level quantum chemical calculations play the important role, which provide the probable structures and the prediction of their IR spectra. Throughout this chapter, we use this experimental and theoretical joint approach to analyze the UV and IR spectra, leading to the determination of the structure of the supramolecules

2. Experimental methods

Figure 1(a) shows the experimental setup of the supersonic beam and laser system. The supersonic jet of the functional molecules is generated by an adiabatic expansion of the gaseous mixture of the sample (host functional molecule) and guest species into the vacuum chamber. The adiabatic expansion generates internally cold gaseous molecules and complexes, with most of them populated in the zero-point vibrational level. However, the cooling during the expansion occurs under non-equilibrium conditions so that several conformers may coexist in the jet. We used a home-built high temperature pulsed nozzle (inset) to generate jet-cooled amino acids, CEs, C4A and their complexes with guest molecules. The pulsed nozzle consists of a commercially available solenoid valve and a sample housing made of polyimide resin. The sample housing which contains sample powder is attached to the head of the commercially available pulse valve and the housing is heated to 140-160 °C to evaporate the samples (Ebata, 2009). The housing has a 1mm orifice at the exit. The opening of the poppet located in the pulse valve is controlled externally to inject the sample gas, which is synchronized with the pulsed lasers. The gaseous mixture of the sample and guest species, premixed with helium or neon carrier gas at a total pressure of 2 bar, is expanded into the vacuum chamber through the orifice. The molecules are internally cooled (T_{rot} = ~10 K) by an adiabatic expansion, and we obtain supersonic free jet. By introducing a skimmer at the downstream of the free jet, a supersonic beam is obtained.

(a)   (b)

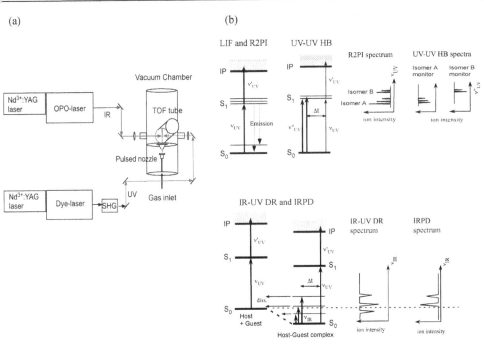

Fig. 1. (a) Experimental setup of the supersonic beam and IR-UV DR spectroscopy. (b) Laser spectroscopic methods used: (upper trace) UV spectroscopy and (lower trace) IR spectroscopy.

We apply several laser spectroscopic methods to obtain the electronic and vibrational spectra of the species generated in the supersonic jet. For the measurement of the electronic spectrum we apply LIF and mass-resolved R2PI [left panel of Figure 1(b)] spectroscopy. For the measurement of LIF spectrum, a tunable nanosecond UV laser pulse crosses the free jet at 20 mm downstream of the nozzle and excites the jet-cooled molecules to the upper electronic state (S_1). The florescence emitted from the molecules is monitored with a photomultiplier tube. By scanning the UV laser frequency while monitoring the total florescence, an LIF spectrum corresponding to the S_1-S_0 spectrum is obtained. For the R2PI measurement, the UV laser ionizes the molecules by one-photon resonant two-photon ionization scheme via the S_1 state. The generated ions are mass-separated by time-of-flight (TOF) tube and detected by a channeltron. The S_1-S_0 UV spectrum is obtained by scanning the UV frequency while monitoring the mass selected ions by a channeltron detector.

The discrimination of the different species in the electronic spectra is carried out by UV-UV HB spectroscopy [right panel of Figure 1(b)] (Ebata, 1998). In this method, two UV laser beams, namely "pump" and "probe" beams, are used. The pump laser pulse crosses the jet at 10 mm downstream of the nozzle, and the probe laser pulse crosses the jet 20 mm downstream of the nozzle. Here, the pump laser light is introduced ~ 4 µs prior to the probe laser light, corresponding to the 10 mm distance of the jet. The frequency of the probe UV laser is fixed to a band of a specific species and that of the pump UV laser is scanned. When the pump laser frequency is resonant to a transition of the monitored species, these species is excited to the upper state resulting in the depletion of the fluorescence or the ion signal monitored by the probe laser light. Thus, the electronic spectrum of the monitored species is obtained as a function of the pump UV frequency.

For the measurement of the IR spectrum of a specific complex we apply IR-UV DR spectroscopy [lower panel of Figure 1(b)] (Brutschy, 2000; Ebata, 2009; Ebata et al., 1998; Tanabe et al., 1993; Zwier, 1996). The principle of this technique is very similar to UV-UV HB spectroscopy, except we use a tunable IR laser light for the pump laser. The IR and UV laser lights are spatially overlapped in the vacuum chamber. The IR laser is introduced ~100 ns prior to the probe UV laser light and its frequency is scanned. Depletion of the monitored signal occurs when the IR frequency is resonant to a vibrational transition of the monitored species and the IR spectrum is obtained as a depletion spectrum. For the complexes of C4A we apply IR photodissociation (IRPD) spectroscopy for obtaining the binding energy [lower panel of Figure 1(b)] (Hontama et al., 2010). In IRPD spectroscopy, the probe UV frequency is fixed to the electronic transition of C4A. When the absorbed IR energy is larger than the binding energy of the C4A complex, the complex dissociates to produce the C4A fragment. Thus, by scanning the IR frequency while monitoring C4A fragment, we obtain the IRPD spectrum. By comparing IRPD and IR-UV DR spectra, we obtain the threshold to generate the C4A fragment which is equal to the binding energy of the complex.

3. Theoretical methods

As was described in the introduction, quantum chemical calculation plays an important role to determine the structures of functional molecules as well as the binding energies. Most of the calculations are carried out by density functional theory (DFT) calculations with the B3LYP or M05-2X functional and the 6-31+G* basis set using the GAUSSIAN 09 program package (Frisch et al., 2009). For the complexes of C4A, in addition to the DFT calculation, a higher level quantum chemical calculation is performed to optimize the structures and obtain accurate binding energies of the complexes. These include the second order Moller-Plesset (MP2) level of theory (Møller & Plesset, 1934) and the family of augmented correlation consistent basis sets of Dunning and co-workers (Dunning, 1989; Kendall et al., 1992) up to quadruple-zeta quality, aug-cc-pVnZ (n = D, T, Q). The MP2/aug-cc-pVDZ optimal geometries were used for single point calculations with the larger basis sets up to aug-cc-pVQZ. For the detail of the calculation, see ref. (Hontama et al., 2010).

The energies of the optimized structures were corrected by zero-point vibrational energy. The harmonic vibrational frequencies were scaled by the factors of 0.97 and 0.95 for the OH and CH stretching vibrations, respectively, in order to compare with the experimentally measured ones. The S_1–S_0 electronic transition energies were calculated using time dependent density functional theory (TD-DFT) with the same functional and basis set.

4. Conformation and hydrated structures of amino acids

4.1 L-Phenylalanine (L-Phe)

The three-dimensional structures of peptides and proteins and their dynamics are to a large extent governed by the conformational profiles of amino acids constituting them. In particular, competition between intra- and inter-molecular hydrogen-bonded interactions plays a vital role in determining the structure of the proteins in solution. In this section, we investigate the conformation of amino acids in bare form and the structures of the hydrated complexes. The conformational landscape of isolated L-Phenylalanine(L-Phe) (Scheme 1) has been examined in detail by several groups (Ebata et al., 2006; Hashimoto et al., 2006; Robertson & Simons, 2001; Snoek et al., 2000; Von Helden et al., 2008). The key point of these studies is that different conformers and isomers exhibit different UV transition energies.

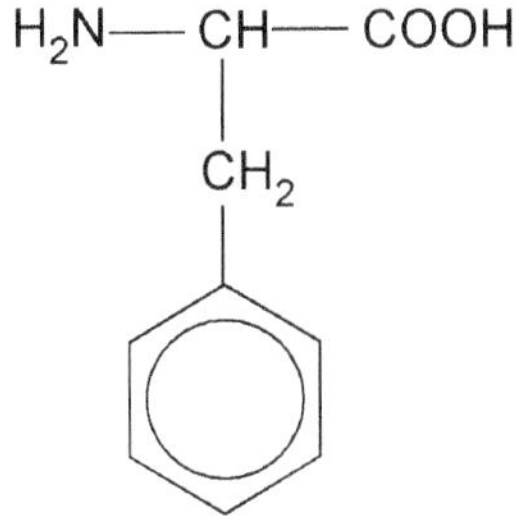

$$H_2N \!-\!\! CH \!-\!\! COOH$$
$$\mid$$
$$CH_2$$

Scheme 1. L-Phenyl alanine

Figure 2(a) shows the LIF and R2PI spectra of jet-cooled L-Phe in the origin region of the S_1-S_0 transition. The inset shows the UV absorption spectrum of L-Phe in solution. There are several peaks in the band origin region, and Figure 2(b) shows the IR-UV DR spectra obtained by monitoring the marked bands (**A, B, C, D, E** and **X**) in the LIF spectrum. Other peaks are assigned to the vibronic bands of these conformers (Snoek et al., 2000). The labeling for each band is adopted from Ref. (Snoek et al., 2000). As seen in Figure. 2(b), their IR spectra are different from each other, and it is concluded that there are at least six conformers of L-Phe in the jet. The conformers **A, C, D,** and **E,** have the free OH stretching vibration of carboxyl OH group at ~3590 cm^{-1}, while the conformers **B** and **X** exhibit largely red-shifted OH stretch at 3280 and 3240 cm^{-1}, respectively. These large red-shifts are due to the intramolecular hydrogen(H)-bonding between the carboxyl OH and the amine nitrogen. The stick diagrams in Figure 2(b) are the calculated IR spectra for the possible conformers for the species **A, B, C, D, E** and **X,** whose structures are shown in Figure 3. In the figure, the number in the parentheses are the zero-point-energy corrected relative energy (kJ/mol) obtained at MP2/6-31+G* level (Ebata et al., 2006).

The structural information gives the conformer dependent photo-physics of L-Phe. By comparing the relative band intensities between the LIF and the R2PI spectra, it is obvious that the relative intensities are quite different between them. Especially the intensity of band **X** is quite weak in the LIF spectrum compared to other bands, while its intensity is comparable to other bands in the R2P1 spectrum. The LIF intensity is proportional to the product of the S_1-S_0 absorption cross section and the fluorescence quantum yield, while the R2PI band intensity is roughly proportional to the absorption cross section. Thus, it is suggested that fluorescence quantum yield of the species associated to band **X** is much smaller than other conformers. Actually, the florescence lifetime of the species **X** is 29 ns, while those of other bands are 70~90 ns (Hashimoto et al., 2006). Thus, it is concluded that S_1 lifetime of L-Phe is quite different for different conformer even in the bare molecule. The short lifetime of conformer **X** having the intramolecular H-bonding may be ascribed by the fast internal conversion or intersystem crossing to the nearby $n\pi^*$ state.

As was described above, there are at least six conformers in the jet, which are classified to the intramolecular H-bonded group and to the non-H-bonded groups. In this section, we investigate how L-Phe changes its conformation to form the complex with water molecules. Figure 4 shows the LIF spectrum of jet-cooled L-Phe measured under different partial pressure

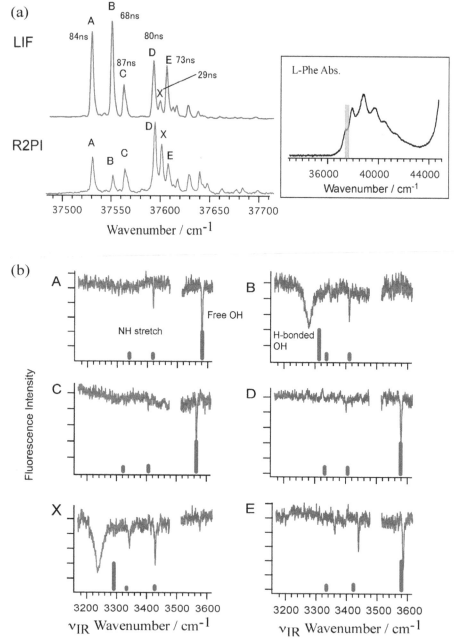

Fig. 2. (a) LIF and R2PI spectra of jet-cooled L-Phe. (b) IR-UV double resonance spectra of L-Phe for the bands marked in the LIF spectrum. Stick diagrams are the calculated IR spectrum of the corresponding conformers shown in Figure. 3. Inset is the UV absorption spectrum of L-Phe in solution. Figure adapted from Ref. (Ebata, 2009).

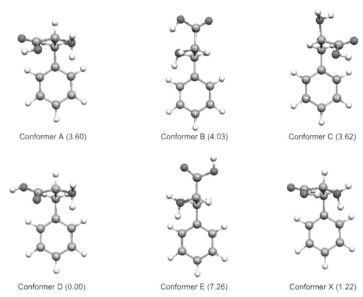

Fig. 3. Optimized structures of L-Phe at the MP2/6-31+G* level. Figure adapted from Ref. (Ebata et al., 2006).

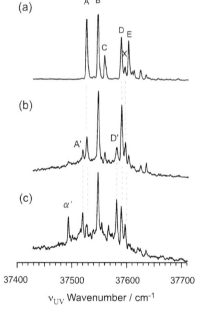

Fig. 4. LIF spectrum of jet-cooled L-Phe and L-Phe-$(H_2O)_n$ obtained under different water vapor pressure condition; (a) without water vapor. (b) water vapor at 0 °C. (c) water vapor at room temperature. He gas at a total pressure of 2.5 bar was used as a carrier gas. Figure adapted from (ref. Ebata et al., 2006)

of water vapor in the He expansion (Ebata et al., 2006). Here, the partial pressure of water vapor increases in the order from (a) to (c). The spectra of Figures 4(b) and (c) are measured at higher sensitivity than that of Figures 4(b), and are normalized by the intensity of band **B**. As seen in Figures 4(b) and (c), the band intensities of the conformers **A, D** and **E** become weaker than the band **B**, and new bands, **A', D'**, and **α'** appear at 37520, 37582 and 37492 cm^{-1}, respectively. The conformers **A, D** and **E** are the non-H-bonded open conformers and it is seen that water molecules easily form H-bonding to the open-type conformers. On the other hand, it seems difficult for the water molecules to be incorporated in the closed conformer by breaking the intramolecular H-bond. The band **A', D'** and **α'** were assigned to L-Phe-(H$_2$O)$_{1,2}$ by mass-selected two-color R2PI measurement (Lee et al., 2002), though their structures were not revealed at that time. It is clear that IR-UV DR spectroscopy unambiguously revealed their structures.

The IR-UV DR spectra observed by fixing UV frequencies to bands **A', D'** and **α'** are shown in the upper panel of Figures 5(a)-(c). In the IR-UV DR spectrum of band **A'**, which is assigned to L-Phe-(H$_2$O)$_1$, there are two intense bands at 3246 and 3506 cm^{-1}, and three sharp bands at 3040, 3421 and 3724 cm^{-1}. Though it is *a priori* not clear whether L-Phe retains a similar conformation under the formation of the H-bonding with water, it is reasonable to assume that the isomer **A'** has the L-Phe conformation similar to conformer **A**, because the intensity of band **A'** increases parallel to the decrease of band **A**, and the small red-shift in the electronic transition of band **A'** with respect to band **A** (7 cm^{-1}) is also observed in other monohydrated aromatic acids. Thus, the structures of L-Phe-(H$_2$O)$_1$ were calculated with L-Phe part forming conformer A. Three H-bonding sites are possible for L-Phe; the carboxyl group, the amino group and the phenyl group. Among them, it was found that the H-bonding to the carboxyl group results in the most stable isomer, and the calculated IR spectrum reproduces very well the observed one. Figure 5(a) also shows the most stable isomer (Aw1-I) and the calculated IR spectrum as a stick diagram. In this isomer, the water forms a cyclic H-bond with the carboxyl group. The intense and broad bands at 3246 and 3506 cm^{-1} are the H-bonded carboxyl OH and the water donor OH stretch bands, respectively, within the cyclic H-bonding network. The sharp bands at 3421 and 3724 cm^{-1} are assigned to the asymmetric NH$_2$ stretch of amino group and free OH stretch of water, respectively. The IR spectrum of band **D'** (Figure 5(b)) is similar to that of band **A'**, because the red-shift of band **D'** from band **D** in the electronic transition, that is 8.5 cm^{-1}, is close to that of band **A'** from band **A**. Thus, the structure of L-Phe-(H$_2$O)$_1$ was calculated with L-Phe forming conformer **D**. It was found that the isomer in which the water molecule forms a cyclic H-bond with the carboxyl group (Dw1-I) is the most stable structure and the IR spectrum of Dw1-I well reproduces the observed one.

Upper panel of Figure 5(c) shows the IR spectrum obtained by fixing UV frequency to band **α'**. In the spectrum, a very broad and strong band is seen at 3002 cm^{-1}. Several sharp bands are overlapped with this broad band, which are assigned to the aromatic CH stretch. Other intense peaks are seen at 3330, 3474, 3680 and 3717 cm^{-1}. The two bands at 3680 and 3717 cm^{-1} can be easily assigned to the OH stretch of water molecules free from H-bonding, and the appearance of the two free OH stretch bands suggests species **α'** to be L-Phe-(H$_2$O)$_2$. From the band position in the LIF spectrum, L-Phe part of species **α'** is thought to have a conformation similar to those bands **A** and **A'**. The calculated possible structures of L-Phe-(H$_2$O)$_2$ (Aw2-I) and their IR spectra are shown in Figure 5(c). In this complex, two waters

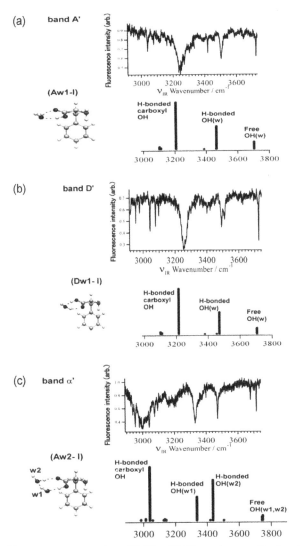

Fig. 5. IR-UV double resonance spectra of L-Phe-$(H_2O)_{n=1,2}$ by monitoring the bands marked in Fig. 4. Stick diagrams are the calculated IR spectra and the corresponding structures obtained at B3LYP/6-31+G* level. Figure adapted from (Ebata et al., 2006)

form cyclic H-bonding with a carboxyl group. A good agreement is seen between the calculated and observed IR spectra of Aw2-I. Thus, it is concluded that band α' is due to the cyclic-form L-Phe-$(H_2O)_2$ of Aw2-I. In Figure 5(c), the band at 3002 cm^{-1} is the H-bonded carboxyl OH stretch. Those at 3330, and 3474 cm^{-1} are the H-bonded OH stretches of two waters in the H-bond ring, and the bands at 3680 and 3717 cm^{-1} are OH stretches free from the H-bond. The weak and broad bands at ~3200 cm^{-1} are due to the overlapped transitions of the aromatic CH stretches. One noticeable point in the IR spectrum of L-Phe-$(H_2O)_2$ is that the frequency shift of the carboxyl OH stretch with respect to bare *L*-Phe is as large as 580 cm^{-1}, which is more than twice the red-shift of the OH stretch of phenol-$(H_2O)_2$ (Watanabe et al., 1996), indicating that the carboxyl OH is considerably weakened under the hydration with two waters. This can be regarded as favorable for the proton transfer to take place in zwitter-ion formation.

4.2 L-tyrosine (L-Tyr)

L-Tyr is obtained by a substitution of the OH group at the para-position of *L*-Phe (Scheme 2). As seen in Figure 6, this substitution causes the number of possible conformers twice of *L*-Phe, arising from the difference of the direction of the OH group with respect to the main frame at the para-position. So, in the figure we label a number 1 or 2 for the classification of the conformers associated to the OH group. Figure 7 shows the LIF spectrum of jet-cooled *L*-Tyr (Inokuchi et al., 2007). The two spectra are obtained at different boxcar gate position to collect the fluorescence; (a) observed by monitoring the total fluorescence, (b) observed by

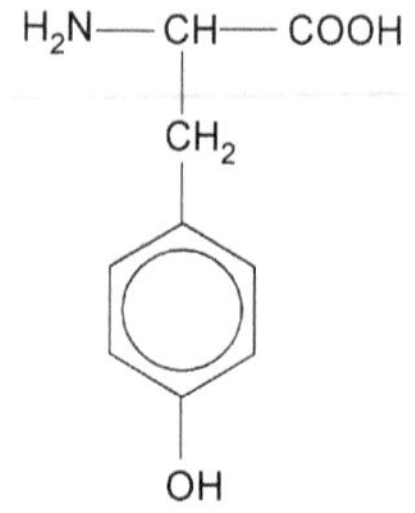

$$H_2N\text{---}CH\text{---}COOH$$
$$|$$
$$CH_2$$

OH

Scheme 2. L-Tyrosine

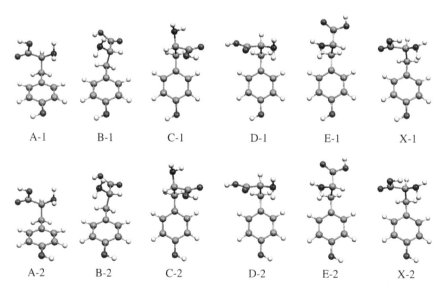

A-1   B-1   C-1   D-1   E-1   X-1

A-2   B-2   C-2   D-2   E-2   X-2

Fig. 6. Optimized structures of L-Tyr at the B3LYP/6-31+G* level. Figure adapted from Ref. (Inokuchi et al., 2007).

monitoring the florescence with the gate position of 12-27 ns after the laser excitation. The extra bands in Figure 7(a) which are not seen in Figure 7(b) are due to Tyramine which are generated by thermal decomposition of L-Tyr. The intensities of these bands increase with an increase of nozzle temperature. In the LIF spectrum, about 20 vibronic bands appeared in the 35450-35750 cm^{-1} region and the assignments of each bands are listed in Table 1.

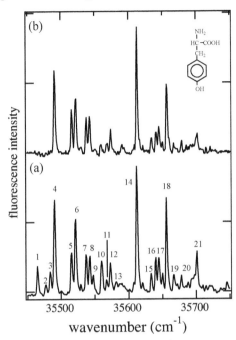

Fig. 7. LIF spectra of jet-cooled L-tyrosine measured with monitoring whole time window of fluorescence (a) and with monitoring a delayed gate position of 12–27 ns from the laser pulse (b). Figure adapted from Ref. (Inokuchi et al., 2007).

UV-UV HB spectroscopy has been applied to discriminate different conformers of L-Tyr, as shown in Figures 8 (Inokuchi et al., 2007). Left panel of the figure displays UV-UV HB spectra obtained by fixing the probe frequencies to bands 4, 6, and 14, respectively. From these spectra, it is concluded that bands 4, 8, and 19 belong to the same conformer. Similarly, the HB spectra indicates that bands 6 and 12, and a shoulder appearing on the lower frequency side of band 21 belong to the same conformer. The HB spectrum of Figure 8d indicates bands 14, 18, and 21 belong to the same species. By the same manner, the discrimination of all the bands can be carried out.

In addition to the discrimination of different species, the HB spectrum shows the features characteristic of the conformation of isomers in the low frequency (torsional) region. Right panel of Figure 8 shows the comparison of the UV-UV HB spectra for different conformers. The HB spectra show rich low frequency vibronic structure, and they can be classified into four groups (k, l, m, n) according to the spectral similarity. The HB spectra of bands 4 and 6 (Figures 8(a) and (b)) show three bands at 39, 52, and 177 cm^{-1} above the origin bands. We classify them into group k. The bands 5 and 7 (Figures 8c and 8d), exhibiting a weak band at 52 cm^{-1}, are classified into group l. The bands 16 and 17 (Figures 8e and 8f), showing two

Band	Position (cm⁻¹)	Fluorescence Lifetime (ns)	Assignment[a]
1	35465	2.8	tyramine
2	35477	2.1	tyramine
3	35484	2.1	tyramine
4 (k_1)	35491	7.1	B
5 (l_1)	35516	5.0	D
6 (k_2)	35522	5.7	B′
7 (l_2)	35537	4.5	D′
8	35543	-	(B)
9	35548	2.4	tyramine
10	35560	2.5	tyramine
11	35567	-	-
12	35573	3.6	(B′)
13	35581	2.1	tyramine
14 (m)	35612	7.9	X, X′
15	35633	4.9	-
16 (n_1)	35640	4.1	E
17 (n_2)	35645	3.5	E′
18	35656	6.6	(X, X′)
19	35667	-	(B)
20	35678	-	-
21	35701	-	(X, X′)

Table 1. Band positions, lifetimes, assignments for vibronic bands in the LIF spectra of L-Tyrosine. [a]Assignments in parentheses show that these bands correlate with origin bands of corresponding conformers.

noticeable bands at 47 and 69 cm⁻¹, are classified into group **n**. The band **14** (Figure 8g) shows two bands at 44 and 89 cm⁻¹, and is classified into group **m**. The low frequency vibrations are assigned to the torsional modes of L-Tyr, such as torsion of the HOOC(H_2N)CH group about the C_α–C_β bond, COOH torsion, and NH_2 torsion. These species belonging to each group are thought to have the very similar conformation of the main frame with each other.

Left panel of Figure 9 shows the comparison of the IR-UV spectra of bands **4**, **6** and **14** with B3LYP calculated ones of the conformers of **B-1** and **B-2** and **X-1** and **X-2** (Inokuchi et al., 2007). The bands **4** and **6** show very similar IR spectra with each other, while that of the band **14** shows the H-bonded OH stretch band at lower position and the asymmetric NH stretch at higher position than those of bands **4** and **6**. The calculated frequencies of the H-bonded OH band of **B-1** and **B-2** (3320 and 3319 cm⁻¹) are higher than those of **X-1** and **X-2** (3293 and 3297 cm⁻¹), and the asymmetric NH bands at 3475 and 3476 cm⁻¹ of **B-1** and **B-2** are lower than those of **X-1** and **X-2** (3491 cm⁻¹). Thus, conformers belonging to bands **4** and **6** are assignable to **B-*n*** and band **14** to **X-*n***. The assignment of bands **4** and **6** to either of **B-1** or **B-2** is not possible at present, and they are labeled as **B** and **B′** (see Table 1). The conformer belonging to band **14** is assigned to " **X, X′** " in Table 1.

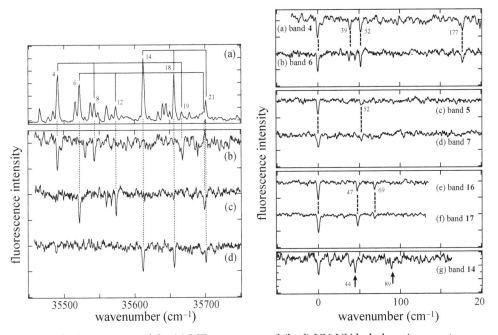

Fig. 8. (Left) Comparison of the (a) LIF spectrum and (b–d) UV-UV hole-burning spectra observed by probing bands **4**, **6**, and **14**. (Right) UV-UV hole-burning spectra for bands **4**, **6**, **5**, **7**, **16**, **17**, and **14**). The abscissa axis represents the energy relative to the transition energy of the origin bands. Figure adapted from Ref. (Inokuchi et al., 2007).

For other species (bands **5**, **7**, **16** and **17**), their conformations can be determined based on the total energies and IR spectra. There are four candidates for them: conformers **A-*n***, **C-*n***, **D-*n***, and **E-*n***. Among them, conformers **A-*n*** are excluded, because the calculated energies of **A-*n*** are much higher than other conformers at B3LYP level of calculation. Conformers **C-*n***, **D-*n***, and **E-*n*** have total energies similar with each other, and are assignable to bands **5**, **7**, **16** and **17** in terms of the total energy. As seen in Figure 6, only conformers **C-*n*** have a OH•••π hydrogen bond between the COOH group and the aromatic ring. A sign of the OH•••π hydrogen bond in conformers **C-*n*** emerges in IR spectra calculated. The right panel of Figure 9 shows comparison of the IR-UV DR for bands **5**, **7**, **16**, and **17** with calculated IR spectra for **D-*n***, **E-*n***, and **C-*n*** (Inokuchi et al., 2007). The OH stretching vibration of the COOH group of conformers **C-*n*** is calculated to be 3569 cm⁻¹, which is lower than those of conformers **D-*n*** and **E-*n*** (3582 and 3583 cm⁻¹, respectively). In the IR-UV DR spectra of *L*-Tyr, bands **5**, **7**, and bands **16**, **17** exhibit the OH stretching band of the COOH group around 3582 cm⁻¹, indicating that these groups do not have any hydrogen bond like the OH•••π bond. Therefore, possible structures for these species are conformers **D-*n*** and **E-*n***. This result is supported by the experimental results of *L*-Phe; conformer C of *L*-Phe in Figure 2 has the carboxyl OH band at 3567 cm⁻¹, which is lower than those of conformer D (3579 cm⁻¹). For the assignment of groups l and n to conformers **D-*n*** or **E-*n***, there is no unambiguous evidence for the assignment in the present stage. However, it is possible that group l is assigned to conformers **D-*n***, and that group n is to **E-*n***, from the relative band intensities of the LIF spectrum and the calculated total

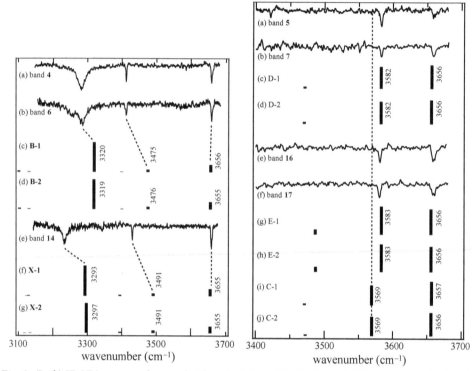

Fig. 9. (Left) IR-UV spectra observed at bands **4**, **6**, and **14** (a, b, and e). IR spectra calculated for conformers B-1, B-2, X-1, and X-2 (c, d, f, and g). (Right) IR-UV spectra observed at bands **5**, **7**, **16**, and **17** (a, b, e, and f). IR spectra calculated for conformers D-1, D-2, E-1, E-2, C-1, and C-2 (c, d, g-j). The dotted line is drawn at 3569 cm^{-1}. Figure adapted from Ref. (Inokuchi et al., 2007).

energies of the conformers. As seen in Figure. 7, bands of group l (bands **5** and **7**) are stronger than those of group n (bands **16** and **17**); the relative intensity is 1:0.75 between groups l and n. Assuming Boltzmann distribution at 363 K, which is the temperature of the sample source, and using the total energy calculated for conformers D-n and E-n, the relative population of conformers **D-n** and **E-n** is estimated at 1:0.65, comparable to the result derived from the band intensity. In addition, the order of the band position in the electronic spectra is **B, D, X**, and **E** from lower to higher frequency in the case of L-Phe. If this order is kept for L-Tyr, groups l and n correspond to conformers **D-n** and **E-n**, respectively.

The above studies on L-Phe and L-Tyr demonstrate the combination of supersonic jet-laser spectroscopic study (LIF, UV-UV HB and IR-UV DR) and the theoretical study is quite powerful to solve the complicated problems such as the determination of conformations of flexible molecules.

5. Encapsulation complexes of calix[4]arene

Calixarenes (CAs) are macrocyclic compounds that are well known in host-guest chemistry. Their structures are comprised of cavities that are formed by phenyl rings connected by methylene and OH groups. The OH groups are strongly hydrogen-bonded with each other at the lower rim of the molecular cavity. This cavity functions as a molecular receptor and encapsulates variety of species through non-covalent interactions, forming various clathrates. The structures of the clathrates or complexes are normally studied by vibrational spectroscopy, NMR and X-ray diffraction methods at room temperature (Atwood et al., 1991; Benevelli et al., 1999; Kuzmicz et al., 2002; Molins et al., 1992). However, the thermal energy at room temperature is often equal to the host-guest interaction energy, especially for the complexes with neutral guest species. This yields very broad or complicated spectra arising from the contribution of possible conformers and fluctuation at the given temperature. The interaction with solvent molecules further affect the structure.

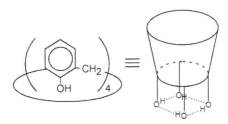

Scheme 3.Calix[4] arene (C4A)

Here, we carry out a supersonic beam-laser spectroscopic study for the complexes of calix[4]arene (C4A, Scheme 3), which is the smallest member of the CAs family. We investigate the structures of the complexes of C4A with variety of neutral guest species, namely Ar, CH_4, N_2, C_2H_2, H_2O, and NH_3 (Ebata et al., 2010; Hontama et al., 2010; Kaneko et al., 2011). These guests are chosen because they can be bound to the C4A host via different interactions, such as dispersion, XH-π hydrogen-bonding and dipole-dipole interactions. We examine how the different nature of the guest/host interaction affects the encapsulation structure, the electronic transition energy and the total binding energy.

Figures 10(a)-(g) show the S_1-S_0 LIF spectra of bare C4A and its complexes (Kaneko et al., 2011). The spectrum of bare C4A shows a sharp 0,0 band at 35357 cm⁻¹. In the higher region, several low frequency vibronic bands are observed. Especially, the vibronic bands at 35500 cm⁻¹ region are stronger than the 0,0 band intensity. C4A has four identical chromophores, phenol, and under the C_4 symmetry their energies are split to ¹A, ¹B and ¹E by exciton coupling (Ebata et al., 2007). Among them, ¹A and ¹E are dipole allowed from the S_0 state. Thus, the band at 35357 cm⁻¹ is assigned to the origin of the ¹A state and the intense bands at 35500 cm⁻¹ are ascribed to the origin of the ¹E state or the vibronic bands appeared by the vibronic coupling between ¹A and ¹E.

The electronic transitions of all the complexes exhibit red-shift with respect to the bare molecule. The inset of Figure 10 shows the plot of the red-shift vs the polarizability of the guest species. For rare gas atoms as guests, a smooth relationship is obtained between the red-shift and the polarizability. Slight deviation for N_2, CH_4 and C_2H_2 from this relationship may be due to the anisotropy of their polarizabilities. For the complexes with H_2O and NH_3, the red-shifts are ~200 cm⁻¹, which is roughly 4 times larger than those of the complexes having the similar polarizabilities. Thus, the red-shift is quite dependent on the type of the guest species.

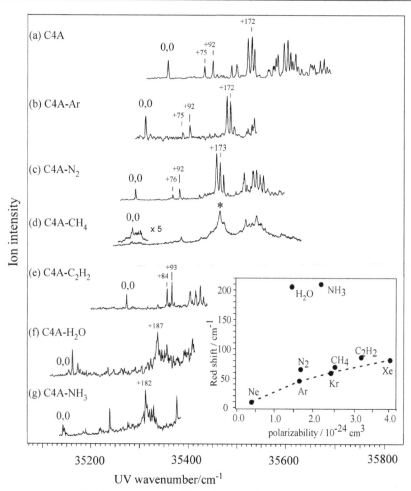

Fig. 10. R2PI spectra of (a) bare C4A and (b)-(g) its complexes. Inset shows the plot of the red-shifts of the complexes as a function of the polarizability of the guest species. Figure adapted from ref. (Kaneko et al., 2011).

Figures 11(a) shows the IR-UV DR spectra of bare C4A (Hontama et al., 2010). Figures 11(b) and (e) show the IR-UV DR and IRPD spectrum of C4A-H_2O. The IR spectrum of C4A exhibits a strong and broad OH stretching band centered at 3160 cm^{-1}. This band is red-shifted by ~500 cm^{-1} from the free OH stretch of phenol (3657 cm^{-1}). Thus, the four OH groups are strongly H-bonded with each other in C4A. The weak band at 3040 cm^{-1} is assigned to the CH stretching vibration of the aromatic ring. In the spectrum of the C4A-$(H_2O)_1$ complex, the broad H-bonded OH stretch band appears at the same frequency of bare C4A but its bandwidth is wider. The similarity of the OH stretching frequency between C4A-$(H_2O)_1$ and bare C4A indicates that the OH groups of C4A are not affected by the complexation with H_2O, that is the water molecule is not bound to the OH groups of C4A. In addition to the strong band at 3160 cm^{-1}, the IR-UV DR spectrum exhibits a weak band at

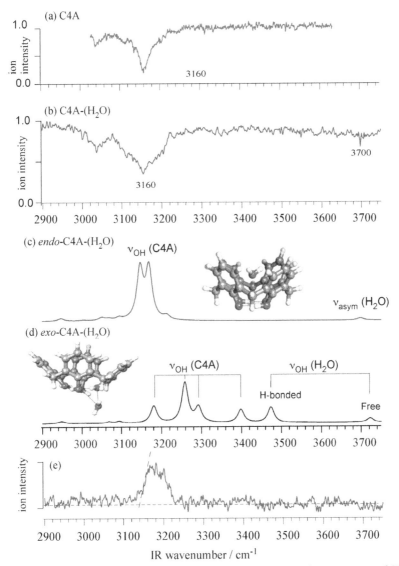

Fig. 11. IR-UV DR spectra of (a) C4A and (b) C4A-(H₂O). Optimized structures and IR spectra of (c) *endo*- (Structure II) and (d) *exo*-form of C4A-(H₂O) (Structure I) obtained at the MP2/aug-cc-pVDZ level of theory. (e) IRPD spectrum of C4A-(H₂O). Figure adapted from Ref.(Hontama et al., 2010).

3700 cm⁻¹, which is assigned to the anti-symmetric OH stretching vibration of the water molecule. This frequency is 56 cm⁻¹ lower than that of gas phase water molecule so that the two OH groups of H₂O are H-bonded in the complex. Figure 11(e) shows the IRPD spectrum of C4A-(H₂O)₁. The spectrum is obtained by scanning the IR laser frequency while monitoring the C4A⁺ signal with a UV frequency fixed near the band origin of C4A. When we compare the IRPD and IR-UV DR spectra we see a sharp cutoff at 3140 ± 20 cm^{-1} in the IRPD spectrum, although the C4A-$(H_2O)_1$ complex shows the IR absorption in the IR-UV DR spectrum. Therefore this cut off means C4A-$(H_2O)_1$ does not dissociate below 3140 ± 20 cm^{-1}, corresponding to the C4A-$(H_2O)_1$ →C4A + H_2O dissociation energy.

The optimal structures of the C4A-$(H_2O)_1$ complex is obtained at MP2/aug-cc-pVDZ levels of theory. The optimization predicts an *endo*-isomer structure, as a global minimum and a *exo*-conformer (Structure I) which are shown in Figures 11(c) and (d). In the *exo*-isomer (Structure I) the water molecule is inserted into and enlarges the ring H-bonding network originally formed by the four OH groups of C4A. The resulting five OH homodromic ring is consistent with the network having the largest cooperativity. In this structure, the H-bonding network of the OH groups in C4A is largely distorted by the insertion of the water molecule. As a result, the calculated IR spectrum of the *exo*-isomer (Structure I) predicts widely distributed OH stretching bands as shown in Figure 11(d). In contrast, the IR spectrum of the global minimum *endo*-isomer (Structure II) (Figure 11(c)) is much simpler, attesting to the minimal distortion of the C4A moiety in the C4A-$(H_2O)_1$ complex. This spectral pattern well reproduces the observed IR-UV DR spectrum of Figure 11(b). The degenerate OH stretching bands of C4A at 3160 cm^{-1} are slightly split into two due to the symmetry reduction since the encapsulation of the water molecule lowers the symmetry of C4A. This is the reason why the observed band at 3160 cm^{-1} is broader than that of bare C4A. The band at 3700 cm^{-1} is assigned to the anti-symmetric OH stretching vibration (v_3) of the encapsulated water. In the *endo*-isomer (Structure II) the two OH groups of the water molecule are bound to two phenyl rings in a bidentate manner and the oxygen atom of the water molecule is facing towards the rim of C4A. This arrangement is thought to originate from the dipole-dipole interactions between C4A and the water molecule. The dipole moment of C4A is 2.37 Debye, oriented along the C_4 axis and pointing upward, and that of the water molecule is 1.855 Debye (Lide, 2001). Thus, the oxygen atom of the water molecule in the cavity prefers to be oriented towards the rim of C4A in order to maximize the dipole-dipole interaction between the two fragments. As was described, the asymmetric OH stretching frequency of the encapsulated water molecule is 56 cm^{-1} lower than that of gas phase water. This lower frequency shift indicates both the OH groups of the water in the cavity are bound to the phenyl group by the OH-π H-bonding. This synergy between the two OH-π H-bonding and dipole-dipole interactions results in the large stabilization energy of "Structure II" with respect to the H-bonded *exo*-isomer (Structure I).

The same synergy effect of the XH-π H-bonding and dipole-dipole interaction is expected for the C4A-$(NH_3)_1$ complex. Here the ammonia molecule has a dipole moment of 1.472 Debye (Lide, 2001). The two stable structures of the C4A-NH_3 complex corresponding to the *endo* and *exo* isomers are shown in Figure 12(a). In the *exo* complex, NH_3 is incorporated in the H-bonding network of the phenolic OH groups, while in the *endo* complex it is encapsulated inside the C4A cavity with the N atom directed to the lower rim. The *endo* complex is calculated to be 34.57 kJ/mol more sable than the *exo* complex at the MP2/aug-cc-pVDZ level. Figure 12(b) shows the comparison of the observed IR-UV DR spectrum with the calculated IR spectra of the two isomers. The calculated IR spectrum of the *endo* complex (Figure 12 (c)) exhibits an intense H-bonded OH band at 3160 cm^{-1}, while the *exo* complex has four distinct OH stretching bands (Figure 12 (d)). It is clear that the calculated IR spectrum of the *endo* isomer reproduces the observed IR spectrum. We can therefore conclude that the observed C4A-NH_3 complex also has the *endo* structure of Figure 13(a). The calculated dipole moments are 3.434 and 3.273 Debye for the C4A-H_2O and C4A-NH_3

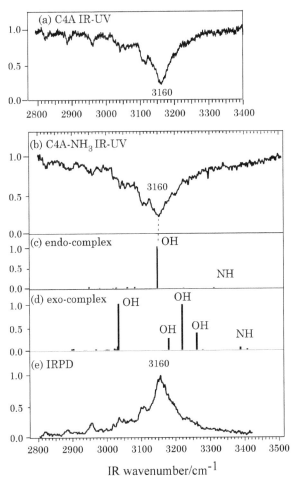

Fig. 12. a) IR-UV DR spectrum of C4A, (b) IR-UV DR spectrum of the C4A-NH₃ complex, (c)-(d) calculated IR spectra of the optimized structures of the *endo*- and *exo*- C4A-NH₃ complexes, (e) IRPD spectrum of C4A-NH₃. Figure adapted from ref. (Kaneko et al., 2011).

complexes, respectively. The binding energy for C4A-NH₃ is experimentally estimated to be less than 2810 cm⁻¹, which is slightly smaller than that of C4A-H₂O (3140 cm⁻¹), a difference that can be attributed to the smaller dipole moment of NH₃ when compared to H₂O as well as the smaller dipole moment of the C4A-NH₃ complex (3.273 D) when compared to that of the C4A-H₂O complex (3.434 D).

Figures 13(b)-(d) show the C4A complexes with N₂, CH₄, and C₂H₂ determined experimentally and by MP2/aug-cc-pVDZ level calculation (Kaneko et al., 2011). All the complexes are the *endo*-complex structure. For C4A-N₂ and -CH₄, they are bound by the dispersion interaction since the red-shifts of the S₁-S₀ transitions fall into the group of the C4A-rare gas complexes (Figure 10). In the C4A-N₂ complex, Figure 13 (b), N₂ is placed along the four-fold axis of C4A. The stability of this complex is due to the dipole-induced-dipole interaction between C4A and N₂. Because the overall C₄ symmetry is retained in the

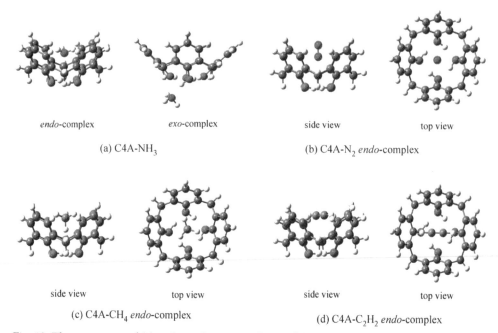

| endo-complex | exo-complex | side view | top view |

(a) C4A-NH₃   (b) C4A-N₂ *endo*-complex

| side view | top view | side view | top view |

(c) C4A-CH₄ *endo*-complex   (d) C4A-C₂H₂ *endo*-complex

Fig. 13. The structures of (a) *endo*- and *exo*-complexes of C4A-NH₃, (b) *endo*- C4A-N₂ complex, (c) *endo*- C4A-CH₄ complex, and (d) *endo*- C4A- C₂H₂ complex. Figure adapted from ref. (Kaneko et al., 2011).

parallel isomer, the observed UV spectrum shows very similar features with that of the bare C4A. The structure of the *endo*-isomer of C4A-CH₄ is shown in Figure 13 (c). In this complex, one CH group is oriented along the 4-fold axis and points to the bottom of the C4A host. For the remaining three other CH groups, two are pointing towards the phenyl rings forming weak CH$-\pi$ hydrogen bonding. In this arrangement, the OH stretching vibrations of the host C4A are unaffected by the complexation with CH₄. Indeed, the observed IR spectrum is very similar to that of bare C4A, a fact that is consistent with this structure. Since the complex does not retain the C_4 point group symmetry, the observed UV spectrum exhibits complicated structure. For the C4A-C₂H₂ *endo*-complex, Figure 13(d), C₂H₂ is located perpendicular to the four-fold axis of C4A so that the two CH groups form CH-π H-bonding with the phenyl groups. TheC4A host is distorted and the overall symmetry is reduced to C_2. This causes the degenerate OH stretching band of C4A to split into two bands, which is also identified in the observed IRPD spectrum.

6. Conformation and recognition of guest species of the crown ether complexes

Crown ethers (CEs) are cyclic ethers built with several oxyethylene (-C-C-O-) units (scheme 4). Applications of crown ethers as molecular receptors, metal cation extraction agents, fluoroionophores and phase transfer catalytic media have been described in a number of studies in the literature (Gokel, 1991; Izatt et al., 1969; Pedersen, 1967; Pedersen &

Frensdorff, 1972). One of the important aspects of the host/guest molecular systems is the selectivity in the encapsulation of guest species. There are two key factors controlling the selectivity: the size and the flexibility of the host cavity. First, if the cavity size of the host molecule fits the size of the guest species, the host shows an efficient selectivity for the encapsulation of the particular species. For example, in solution the 18-crown-6-ether (18C6) forms an exceptionally stable 1:1 complex with K^+ among alkali metal cations, because 18C6 forms a ring conformation of D_{3d} symmetry and the size of its cavity is comparable to the size of the spherical K^+. However, in the gas phase, 18C6 as well as 12-crown-4 (12C4) and 15-crown-5 (15C5) show the largest binding energy to Li^+ not to K^+ among the alkali-metal cations (Anderson et al., 2003; Armentrout, 1999; Glendening et al., 1994; More et al., 1999; Peiris et al., 1996). Very recently, it is reported that this largest binding energy of L^+-18C6 in the gas phase is due to the structure in which 18C6 can distort its ether frame to shorten the ether oxygen distance and maximize the biding energy (Inokuchi et al., 2011). Thus, the selectivity is substantially affected by the solvent molecules as well, and a stepwise study starting from the isolated molecule to micro-solvated complexes is essential. Molecular complexes provide an ideal environment for the precise study of the micro-solvated effects under solvent-controlled conditions. As was described above, the flexibility is the key point of crown ethers and they are able to adjust their conformations to fit the size and the shape of the guest species. We show how the structural flexibility affects the dynamics of the encapsulation for two CEs;dibenzo-18-crown-6-ether (DB18C6) and benzo-18-crown-6-ether (B18C6) and their complexes with various neutral molecules (Kusaka et al., 2007, 2008, 2009, 2011)

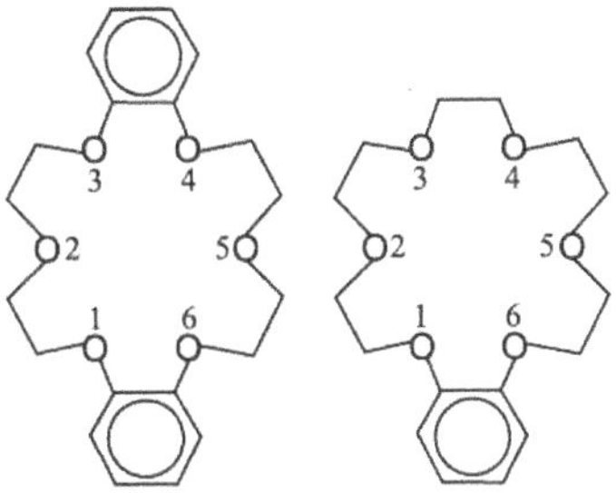

Scheme 4. Dibenzo-18-Crown-6 (DB18C6, left) and Benzo-18-Crown-6(B18C6, right)

6.1 DB18C6 and its complexes

Figure 14(a) shows the LIF spectrum of DB18C6 cooled in a supersonic jet (Kusaka et al., 2008). Figures 14 (b)-(h) show the UV-UV HB spectra measured by monitoring bands m1, m2, a, and c-f, respectively. From the UV-UV HB and mass selected R2PI measurements, it is confirmed that the bands m1 and m2 belong to the different conformers of DB18C6, and bands a-f to the DB18C6-$(H_2O)_{n=1-4}$ complexes. There are three noticeable points in the LIF spectrum. First for bare DB18C6, m2 is the major species from the relative intensity. Second, all the hydrated complexes show blue-shifted band origins with respect to bare molecule. This means that DB18C6 plays the role of the acceptor of the H-bonding. Finally, the band origins of m2 and a exhibit a 5 cm^{-1} splitting, which is descried as the exciton splitting of the two benzene chromophores in the same environment. In the LIF spectrum, the assignments of each bands are listed in Table 2.

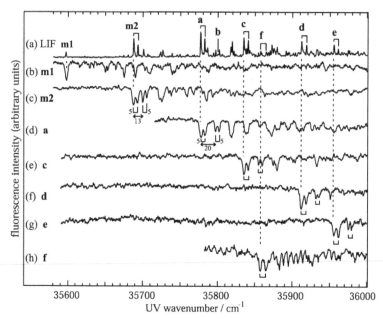

Fig. 14. (a) LIF spectrum of jet-cooled DB18C6 and its hydrated complexes. (b)-(h) UV-UV HB spectra measured by monitoring bands **m1**, **m2**, **a**, and **c-f** in the LIF spectrum, respectively. The numbers in (c) and (d) show the energy interval (cm⁻¹) in the corresponding regions. Figure adapted from Ref. (Kusaka et al., 2008).

Position / cm⁻¹	Label	Size	Assignment
35597	**m1**	DB18C6	IV
35688	**m2**		II
35777	**a**	DB18C6-$(H_2O)_1$	1W-1
35800	**b**		1W-2
35835	**c**	DB18C6-$(H_2O)_2$	2W-1
35858	**f**	DB18C6-$(H_2O)_4$	4W-2
35912	**d**	DB18C6-$(H_2O)_3$	3W-1
35955	**e**	DB18C6-$(H_2O)_4$	4W-1

Table 2. Band positions of origins for bare DB18C6 and the hydrated complexes (See Figure 16)

The IR-UV DR spectra in the CH stretching region for species **m1**, **m2**, and **a** are shown in Figure 15(b) (Kusaka et al., 2011). In the spectra, the bands in the 2800-3000 cm⁻¹ region are the CH stretching vibrations of the methylene groups and those in the 3000-3100 cm⁻¹ region to the CH stretching vibrations of benzene rings. In the methylene CH stretching region, the spectral patterns of **m1** and **m2** are similar with each other though they are the different conformers. On the other hand, while species **a** shows a quite different spectrum. For example, **m1** and **m2** show a strong band at 2950 cm⁻¹, while species **a** does not. Instead, species **a** shows a strong band at 2830 cm⁻¹ and weak one at 2800 cm⁻¹. Since IR spectra in the CH stretching region can reflect the conformation of crown ethers, the IR spectra in Figure 15(b) indicate that **m1** and **m2** have a structure similar with each other, while the DB18C6 conformation of species **a**, DB18C6-(H₂O)₁, is quite different from them.

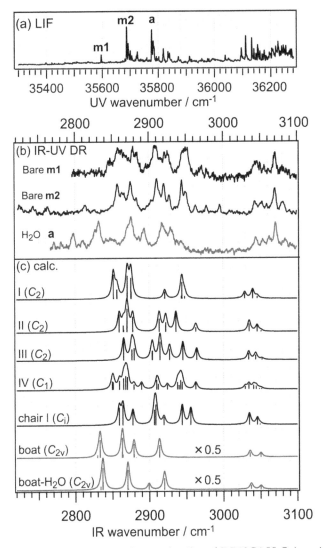

Fig. 15. (a) LIF spectrum of bare DB18C6 (**m1** and **m2**) and DB18C6-H₂O (species **a**). (b) IR-UV DR spectra of **m1**, **m2**, and **a**. (c) Calculated IR spectra of optimized bare DB18C6 and DB18C6-H₂O at M05-2X/6-31+G* level. The optimized geometries are shown in Figure 16. Figure adapted from Ref. (Kusaka et al., 2011).

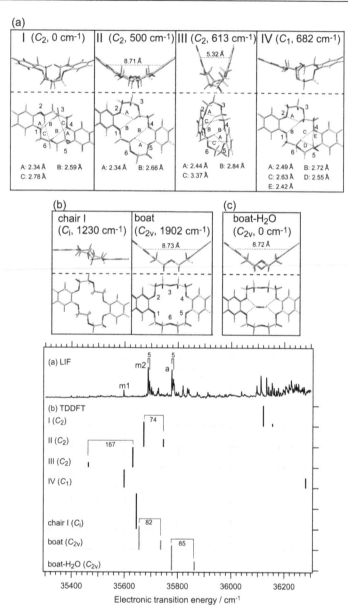

Fig. 16. (Upper) Optimized structures of bare DB18C6 and DB18C6-H$_2$O. (a) four most stable structures of bare DB18C6, (b) bare DB18C6 (chair I and boat), and (c) DB18C6-H$_2$O (boat-H$_2$O). Relative energies with respect to the most stable structure are displayed in cm^{-1} unit. The distances of CH$\cdots$O, CH$\cdots\pi$ and $\pi\cdots\pi$ are also indicated. (Lower) (a) LIF spectrum of bare DB18C6 and DB18C6-H$_2$O. (b) S$_1$-S$_0$ and S$_2$-S$_0$ electronic transition energies (bar graph) obtained by TDDFT calculations at the M05-2X/6-31+G* level. Figure adapted from Ref. (Kusaka et al., 2011).

The determination of the structures of **m1**, **m2**, and **a**, is carried out by the comparison of the observed IR spectra with the ones of the optimized structures obtained by quantum chemical calculations. Figure 16 shows the four most stable conformers of bare DB18C6 obtained at M05-2X/6-31+G* level (Kusaka et al., 2011). The energies (cm^{-1}) relative to the most stable conformer are displayed in the parentheses with the symmetry. In the figure, higher energy conformers; "chair I" and "boat" conformers, are also shown as the candidates for **m1** and **m2**. Figure 16 also shows the boat form DB18C6-(H$_2$O)$_1$, which is the most stable isomer of the 1:1 complex. In the structures in Figure 16, the crown frame of conformers I, II, III, and IV is fixed by the CH$\cdots$O and $\pi\cdots\pi$ interactions, while such interactions seem very weak in the boat conformer.

Figure 15(c) shows the calculated IR spectra of the conformers of bare DB18C6 and boat-DB18C6-H$_2$O complex shown in Figure 16. The calculated IR spectrum of the boat conformer is quite different from those of the other conformers; it exhibits fewer IR bands than those of the other conformers because of its higher symmetry (C$_{2v}$). The calculated IR spectrum of boat DB18C6-H$_2$O well reproduces that of the boat conformer. The key for identifying the boat conformation is the appearance of bands at 2835 cm^{-1} and no band at $\sim$ 2950 cm^{-1}. Thus, from the energetic and the similarity of the IR spectrum, it is concluded that the structure of species **a** is the boat-DB18C6-H$_2$O of Figure 16(c). We also see that the calculated IR spectrum of boat DB18C6-H$_2$O is very similar with bare boat DB18C6 as seen in the figure. On the other hand, the IR-UVDR spectra of species **m1** and **m2** (Figure 15 b) are quite different from that of the calculated boat DB18C6. Thus, the initial conformation of bare DB18C6 is not the boat form.

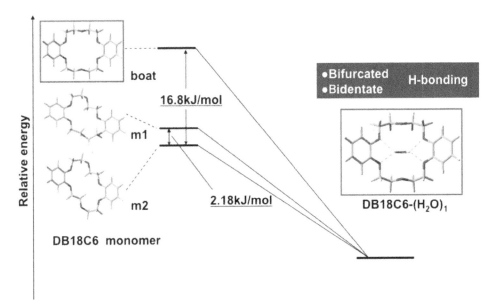

Fig. 17. Energetics of bare DB18C6 and DB18C6-H$_2$O obtained at the M05-2X/6-31+G* level. In bare DB18C6, the boat form is the higher energy conformer, while the boat conformer becomes the lowest energy conformer when it encapsulates a water molecule. In the complex, DB18C6 and H$_2$O are bound via bifurcated and bidentate hydrogen-bonding.

The determination of the structures of species **m1** and **m2** is carried out by the comparison of the observed IR and the electronic spectrum with the theoretically calculated ones. In the IR spectrum of the CH stretching region (Figure 15), we see that the spectra of conformers I-IV resemble each other and the observed IR spectra of species **m1** and **m2**. Thus, it is difficult to determine the structures from the IR spectra. So, we compared the electronic transition energies. Lower panel of Figure 16 shows the comparison of the LIF spectrum and the transitions of the different conformers of DB18C6 obtained by TD-DFT calculation. The calculated transition energies are scaled so that the energy of the boat DB18C6-H_2O fits to the observed band **a**. The split of ~80 cm^{-1} for conformers II and boat is due to the exciton splitting of DB18C6 in which the two chromophores are under the same environment of symmetry. Actually, the splitting is much smaller in the real molecules because the oscillator strength is spread to vibronic bands according to Franck-Condon principle. The observed splitting is 5 cm^{-1} for bands **m2** and **a**. This small exciton splitting indicates that band **m2** is due to the conformer II having the C_2 symmetry. The electronic transitions of other conformers are largely split due to the different environments of the chromophores. Among them, the positions of the conformers IV and Chair I are close to the band **m1**. Since conformer IV is lower energy conformer as shown in Figure 16, we conclude that band **m1** is due to the conformer IV.

The above results show that bare DB18C6 changes its conformer to the boat form to encapsulate the water molecule in this cavity, though the boat form is not the most stable structure in bare form shown in Figure 17. The boat conformation becomes most stable by the bidentate and the bifurcated hydrogen-bond formation. This conformation change is also found in the DB18C6-NH_3 complex, where the bidentate and the bifurcated hydrogen-bond is formed between the host DB18C6 and guest NH_3. On the other hand, when we use CH_3OH or C_2H_2 as a guest species such a conformation change does not occur since they can not form the bidentate and the bifurcated hydrogen-bonding. Thus, we can say that DB18C6 can recognize the difference of the guest species from the difference of the type of the H-bonding.

The hydrogen-bonded structures of DB18C6-$(H_2O)_n$ are revealed from the analysis of the IR-UV DR spectra in the OH stretching vibration, which are shown in the upper panel of Figure 18 (Kusaka et al., 2008). Figures 18(a) and (b) show the IR spectra for bands **a** and **b**, respectively, where the species of band **a** is the major species. From the appearance of two OH bands at different positions, they are due to different isomers of the DB18C6-$(H_2O)_1$ complex. The positions of the two vibrations are red shifted by 77 and 108 cm^{-1} for band **a** , and by 51 and 77 cm^{-1} for band **b** with respect to the vibrations of gas phase water, respectively. This is the evidence that in both complexes forms the bifurcated and bidentate H-bonding. Figure 18(c) shows the IR-UV DR spectrum of species **c**, corresponding to the DB18C6-$(H_2O)_2$ complex. The bands at 3562 and 3623 cm^{-1} are assigned to the symmetric and anti-symmetric OH stretching vibrations of a bidentate water molecule, respectively. Their positions are red shifted by 18 and 25 cm^{-1} from those of band **a**, meaning the O atom of the bidentate water acts as an acceptor for the second water molecule. The bands at 3401 and 3716 cm^{-1} can be assigned to the singly H-bonded and free OH stretching vibrations of the second water molecule, respectively. The IR-UV DR spectrum of band **d** in Figure 18(d) exhibits six OH stretching bands, suggesting this species is DB18C6-$(H_2O)_3$. In the spectrum the bands are located close to each other and no band appears at the free OH stretching region. Therefore all water molecules form bidentate H-bonds. The six bands are classified to three pairs of the symmetric and anti-symmetric OH stretching vibrations of the bidentate

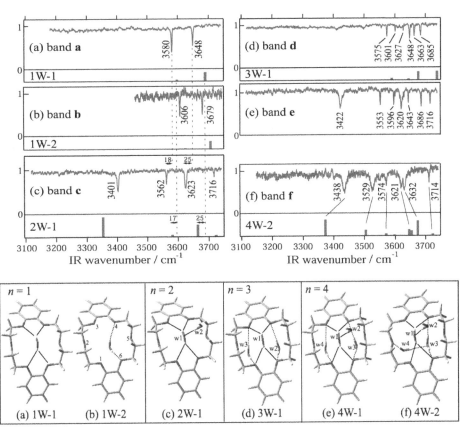

Fig. 18. (Upper) (a)-(f) IR-UV DR spectra of DB18C6-(H$_2$O)$_n$ measured by monitoring bands a-f in the LIF spectrum, respectively. Sticks under the IR-UV DR spectra denote the calculated IR spectra at the optimized structures. (Lower) Geometric features deduced from the analysis of the IR-UV DR spectra in the OH stretching region of species a-f of the DB18C6-(H$_2$O)$_n$ complexes. Figure adapted from Ref. (Kusaka et al., 2008).

water molecules. The lowest pair of the frequencies (3575 and 3648 cm^{-1}) is attributed to a bidentate water molecule bound to the bottom of the boat conformer and the other two pairs arise from the water molecules forming weaker bidentate H-bonds at the opposite (top) side of DB18C6. The spectrum of band e in Figure 18(e) shows seven bands and the band at 3620 cm^{-1} has a shoulder, indicating this species is DB18C6-(H$_2$O)$_4$. The 3422 and 3716 cm^{-1} bands can be assigned to the H-bonded and free OH stretching vibrations of a single-donor water molecule, respectively. The IR spectrum of species e in the 3550-3690 cm^{-1} region is very similar to species d. This suggests that the IR bands of species e can be assigned to the OH stretching bands of three water molecules H-bonded like those in 3W-1. The IR-UV DR spectrum of species f in Figure 18(f) indicates that this species is the isomer of DB18C6-(H$_2$O)$_4$. The 3438 cm^{-1} band can be assigned to the single-donor OH stretching vibration and the band at 3714 cm^{-1} is assigned to the free OH stretching vibration. The 3529 cm^{-1} band is unique to species f. This band is broad and is located on the lower frequency side of the bidentate symmetric OH stretching vibration. Therefore, the 3529 cm^{-1} band cannot be assigned to a bidentate water H-bonded to the ether O atoms.

The lower panel of Figure 18 shows the most probable structures for species **a-f** obtained by DFT calculation and their predicted IR spectra are shown as blue bar graph. The two conformers, 1W-1 and 1W-2, of the DB18C6-(H$_2$O)$_1$ complex, correspond to species **a** and **b**, respectively. In both structures the conformation of DB18C6 is the boat form and the water molecule is H-bonded to the O atoms next to the benzene rings by bidentate H-bonding. However, in 1W-1, the water molecule is H-bonded to DB18C6 by the bifurcated and bidentate manner while in 1W-2 the two OH groups of the water are bonded directly to the O$_4$ and O$_6$ atoms, respectively. The OH stretching frequencies of 1W-1 are lower than those of 1W-2 due to the stronger hydrogen bonds. 2W-1 corresponds to species **c** (DB18C6-(H$_2$O)$_2$), in which the second water molecule (w2) is H-bonded to the O atom of the bidentate water molecule (w1). A noticeable feature of the spectra of species **c** is that the singly H-bonded OH stretching frequency (3401 cm^{-1}) is much lower than that of water molecules forming a normal H-bond. For example, the frequency of the donor OH stretching vibration in the water dimer is 3530 cm^{-1}. This large frequency reduction is due to that the O atom of the bidentate water molecule is highly negatively charged so that the second water molecule (w2) forms a strong H-bond.

3W-1corresponds to species **d** (DB18C6-(H$_2$O)$_3$). In this structure the first water molecule (w1) forms a bidentate H-bond at the bottom side of the boat DB18C6 like 1W-1 and the second (w2) and third (w3) water molecules form two bidentate H-bonds at the opposite side like 1W-2. The 4W-1 represents a hypothetical probable structure for species **e** (DB18C6-(H$_2$O)$_4$). This structure was not obtained as the stable form at the level of calculation used. 4W-2 shows a probable structure for species **f** (DB18C6-(H$_2$O)$_4$). In 4W-2, the first and second water molecules (w1 and w2) construct a bidentate and single-donor H-bonded network like 2W-I, whereas the third water molecule (w3) is bonded to the O$_4$ and O$_6$ atoms like 1W-2 and the fourth water molecule (w4) forms a bridge between an ether O atom (O$_2$) and the O atom of (w3). This type of H-bonding network was also found in the 18-crown-6-ether/water system at the liquid nitrogen temperature. The calculated IR spectrum for 4W-2 well predicts the band at ~3500 cm^{-1}, which is the stretching vibration of (w4) bonded to the O atom of (w3).

6.2 B18C6 and its complexes

Substitution of phenyl group(s) to CE makes the ether ring more rigid because the oxygen atoms adjacent to the benzene ring prefer the planar structure due to the delocalization of the π-electrons of benzene ring. In this sense, B18C6 is more flexible than DB18C6. We investigate how the difference in the flexibility affects its conformation as well as its complexation with water. Upper panel of Figure 19 shows LIF spectra of B16C6 in the band origin region observed without and with adding water vapor, respectively (Kusaka et al., 2009). Bands **M1-M4** and **A-I** are due to bare B18C6 and the B18C6-(H$_2$O)$_n$ complexes, respectively, because the addition of water vapor reduces the intensities of bands **M1-M4** and increases bands **A-I**.

Lower panel of Figure 19 show the observed IR-UV DR spectra for bands **A-I** in the LIF spectrum. The species associated with bands **A-D**, show a pair of the OH stretching bands of the bidentate hydrogen-bonding, indicating different isomers of B18C6-(H$_2$O)$_1$. Species **E**, **F** and **G** are due to B18C6-(H$_2$O)$_2$ complexes. The IR spectra of **E** and **G** show that each of two water molecules forms bidentate hydrogen-bonding to B18C6, while the species **F** shows the

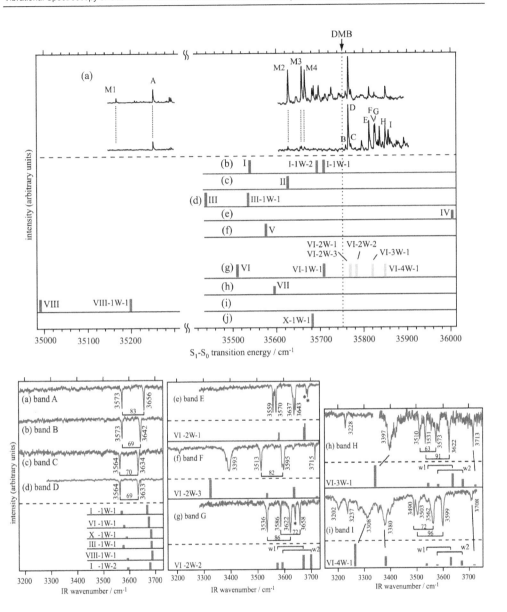

Fig. 19. (Upper) (a) LIF spectra of jet-cooled B18C6 and its hydrated complexes obtained without and with adding water vapor. (b-j) TD-DFT calculated S_1-S_0 transition energies of the structures given in Figure 20. The red, blue, and green bars correspond to bare B18C6, B18C6-$(H_2O)_1$, and B18C6-$(H_2O)_{2-4}$, respectively. (Lower) IR-UV DR spectra (red) of the obtained by monitoring the bands in the LIF spectrum. Also shown are the calculated IR spectra (blue) of the structures given in Figure 20. Figure adapted from Ref. (Kusaka et al., 2009).

IR spectrum similar to that of species **c** of DB18C6-$(H_2O)_2$, Figure 18. The IR spectra of higher size complexes, **H** and **D**, exhibits the two pair of OH stretching bands associate with the bidentate hydrogen-bonding. Species **A** and species **B-D** can be assigned to isomers of B18C6-$(H_2O)_1$ built on **M1** and **M2-M4**, respectively from the similar values of blue-shifts (~100 cm-1) of the complexes. Interestingly, the intensity of band **D** is strongest among the isomers of B18C6-$(H_2O)_1$, which means that species **D** has the structure preferred for the complex formation with the water molecule. Bands **E-I** are assigned to larger hydrated complexes built based on species **D**.

B18C6 (M1 · M4)

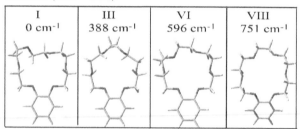

I 0 cm-1	III 388 cm-1	VI 596 cm-1	VIII 751 cm-1

B18C6·H_2O (A · D)

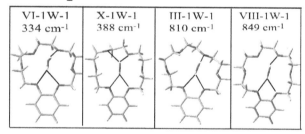

VI-1W-1 334 cm-1	X-1W-1 388 cm-1	III-1W-1 810 cm-1	VIII-1W-1 849 cm-1

B18C6·$(H_2O)_{n=2 \cdot 4}$ (E · I)

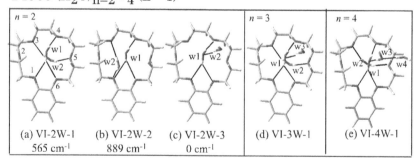

n = 2			n = 3	n = 4
(a) VI-2W-1 565 cm-1	(b) VI-2W-2 889 cm-1	(c) VI-2W-3 0 cm-1	(d) VI-3W-1	(e) VI-4W-1

Fig. 20. (Upper) Most probable structures of B18C6 and B18C6-$(H_2O)_1$. All the calculations are done at B3LYP/6-31+G* level. (Lower) Probable structures of (a-c) B18C6-$(H_2O)_2$, (d) B18C6-$(H_2O)_3$, and (e) B18C6-$(H_2O)_4$ at the B3LYP/6-31+G* level. The numbers shown in cm-1 represent the electronic energies of the isomers relative to that of VI-2W-3. Solid and dotted lines show H-bonding from the front and back sides of B18C6, respectively. Figure adapted from Ref. (Kusaka et al., 2009).

The determination of the structures of bare B18C6 and B18C6-$(H_2O)_1$ is not so straight forward, because of larger number of possible conformers and isomers than those of DB1816. In the calculation, we obtained eight lowest energy conformers for bare B18C6 within the energy of 800 cm^{-1} at B3LYP/6-31+G* level. The IR spectra in the CH stretching region were not helpful. Instead, we determined the structures by two steps; (1) we first examined whether each conformer has the structure which can incorporate H_2O by the bidentate H-bonding. (2) we then compared the TD-DFT calculated electronic transition energy with the observed transition. Figure 20 shows the possible structures for **M1-M4** and **A-D**. Among the B18C6-$(H_2O)_1$ complexes of **A-D**, the intensity of species **D** is strongest, meaning this species with the structure of VI or X is the preferred geometry for the encapsulation of H_2O. We also see that larger size B18C6-$(H_2O)_n$ complexes are grown based on this structure, Figure. 21. It is not clear why B18C6 prefers the conformer VI or X, upon the complex formation with water because neither conformer VI nor X is the most stable structure at the B3LYP/6-31+G* level calculation. However, since the energy differences for different conformers are very small and it is quite possible that higher level of calculation will give more reliable energies. Anyway, similar to the case of DB18C6, there is a preferred structure of B18C6 to incorporate water molecule(s), which is strongly related to the molecular recognition of the host/guest complexes.

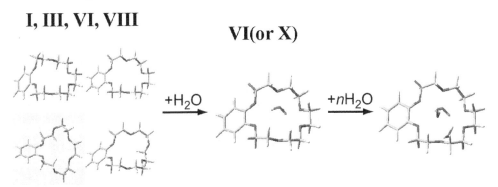

Fig. 21. Conformer preference for the formation of B18C6-$(H_2O)_n$ host-guest complexes. Figure adapted from Ref. (Kusaka et al., 2009).

7. Conclusion

In this chapter, we described our combined study of laser spectroscopic work and theoretical calculation for the determination of the structures of gas phase amino acids and host/guest complexes cooled in supersonic jets. The cooling effect of the supersonic jets enabled to simplify the electronic spectra of those molecules and complexes which are difficult in condensed phase. The double resonance spectroscopic technique, UV-UV hole-burning and IR-UV double resonance, is quite powerful to discriminate difference species including conformers and isomers and to measure the IR spectrum of individual species. For amino acids, it was found that different conformers have different fluorescence lifetimes though its reason is not clear. Also it was found that water molecules prefer the open conformers for the hydration. For the encapsulation complex of calix[4]arene, we showed that it can encapsulate variety of guest species in its cavity by use of difference interaction depending on the guest. In the crown ether systems, a conformer change and preference were observed for forming the inclusion complexes for specific guests. These are the simplest examples of the molecular recognition of the host-guest complex. In near future, this study will be extended to much larger molecular systems by developing the technique of vaporization of nonvolatile large molecule, such as using laser ablation and electro-spray.

8. References

Anderson, J. D.; Paulsen, E. S. & Dearden, D. (2003). Alkali metal binding energies of dibenzo-18-crown-6: experimental and computational results. *International Journal of Mass Spectrometry*, Vol.227, No.1 pp. 63-76, ISSN 1387-3806

Armentrout, P. B. (1999). Cation–ether complexes in the gas phase: thermodynamic insight into molecular recognition. *International Journal of Mass Spectrometry*, Vol.193, No.2-3 pp. 227-240, ISSN 1387-3806

Atwood, J. L.; Hamada, F.; Robinson, K. D.; Orr, G. W. & Vincent, R. L. (1991). X-ray diffraction evidence for aromatic π hydrogen bonding to water. *Nature*, Vol.349, No.6311, pp. 683-684, ISSN 0028-0836

Atwood, J.L.; Barbour L. J. & Jerga A. (2002) Organization of the interior of molecular capsules by hydrogen bonding. *Proceedings of the National Academy of Sciences*, Vol.99, No.8, pp. 4837-4841, ISSN 0027-8424

Benevelli, F.; Kolodziejski, W.; Wozniak, K. & Klinowski, J. (1999). Solid-state NMR studies of alkali metal ion complexes of *p*-tertbutyl-calixarenes. *Chemical Physics Letters*, Vol.308, No.1-2, pp. 65-70, ISSN 0009-2614

Brocos P.; Banquy X.; Díaz-Vergara N.; Pérez-Casas S.; Costas M. & Piñeiro Á. (2010). Similarities and Differences Between Cyclodextrin–Sodium Dodecyl Sulfate Host–Guest Complexes of Different Stoichiometries: Molecular Dynamics Simulations at Several Temperatures. *The Journal of Physical Chemistry B*, Vol.114, No.39, pp. 12455-12467, ISSN 1089-5647

Brutschy, B. (2000). The Structure of Microsolvated Benzene Derivatives and the Role of Aromatic Substituents. *Chemical Reviews*, Vol.100, No.11 pp. 3891-3920, ISSN 0009-2665

Dunning, T. H., Jr. (1989). Gaussian basis sets for use in correlated molecular calculations. I. The atoms boron through neon and hydrogen. *Journal of Chemical Physics*, Vol.90, No.2 pp. 1007-1989, ISSN 0021-9606

Ebata, T. (1998). Population Labelling Spectroscopy, In: *Nonlinear Spectroscopy for Molecular Structure Determination*, Field, R. W., Hirota E., Maier J. P., Tsuchiya, S., (pp. 149-165), Blackwell Science, ISBN 0-632-04217-6, Oxford

Ebata, T.; Fujii, A. & Mikami, N. (1998). Vibrational spectroscopy of small-sized hydrogen-bonded clusters and their ions. *International Reviews in Physical Chemistry*, Vol.17, No.4, pp. 331-361, ISSN 0144-235X

Ebata, T. ; Hashimoto, T.; Ito, T. ; Inokuchi, Y. ; Altunsu, F. ; Brutschy, B. & Tarakeshwar, P. (2006). Hydration profiles of aromatic amino acids: conformations and vibrations of L-phenylalanine-$(H_2O)_n$ clusters. *Physical Chemistry Chemical Physics*, Vol.8, No.41, pp. 4783-4791, ISSN 1463-9076

Ebata, T.; Hodono, Y.; Ito, T. & Inokuchi, Y. (2007). Electronic spectra of jet-cooled calix[4]arene and its van der Waals clusters: Encapsulation of a neutral atom in a molecular bowl. *Journal of Chemical Physics*, Vol.126, No.14 pp. 141101-1-14101-4, ISSN 0021-9606

Ebata, T. (2009). Study on the Structure and Vibrational Dynamics of Functional Molecules and Molecular Clusters by Double Resonance Vibrational Spectroscopy. *Bulletin of the Chemical Society of Japan*, Vol.82, No.2, pp. 127-151, ISSN 0009-2673

Ebata, T.; Hontama, N.; Inokuchi, Y.; Haino, T.; Apra, E. & Xantheas, S. S. (2010). Encapsulation of Ar_n complexes by calix[4]arene: *endo-* vs. *exo-*complexes. *Physical Chemistry Chemical Physics*, Vol.12, No.18, pp. 4569-4579, ISSN 1463-9076

Frisch, M. J.; Trucks, G. W.; Schlegel, H. B.; Scuseria, G. E.; Robb, M. A.; Cheeseman, J. R.; Scalmani, G.; Barone, V.; Mennucci, B.; Petersson, G. A.; Nakatsuji, H.; Caricato, M.; Li, X.; Hratchian, H. P.; Izmaylov, A. F.; Bloino, J.; Zheng, G.; Sonnenberg, J. L.; Hada, M.; Ehara, M.; Toyota, K.; Fukuda, R.; Hasegawa, J.; Ishida, M.; Nakajima, T.; Honda, Y.; Kitao, O.; Nakai, H.; Vreven, T.; Montgomery, J. A., Jr.; Peralta, J. E.; Ogliaro, F.; Bearpark, M.; Heyd, J. J.; Brothers, E.; Kudin, K. N.; Staroverov, V. N.; Kobayashi, R.; Normand, J.; Raghavachari, K.; Rendell, A.; Burant, J. C.; Iyengar, S. S.; Tomasi, J.; Cossi, M.; Rega, N.; Millam, J. M.; Klene, M.; Knox, J. E.; Cross, J. B.; Bakken, V.; Adamo, C.; Jaramillo, J.; Gomperts, R.; Stratmann, R. E.; Yazyev, O.; Austin, A. J.; Cammi, R.; Pomelli, C.; Ochterski, J. W.; Martin, R. L.; Morokuma, K.; Zakrzewski, V. G.; Voth, G. A.; Salvador, P.; Dannenberg, J. J.; Dapprich, S.; Daniels, A. D.; Farkas, O.; Foresman, J. B.; Ortiz, J. V.; Cioslowski, J. & Fox, D. J. (2009). Gaussian 09, revision A.02; Gaussian, Inc.: Wallingford, CT

Glendening, E. D.; Feller, D. & Thompson, M. A. (1994). An Ab Initio Investigation of the Structure and Alkali Metal Cation Selectivity of 18-Crown-6. *Journal of the American Chemical Society*, Vol.116, No.23, pp. 10657-10669, ISSN 0002-7863

Gokel, G. W. (1991). *Crown Ethers and Cryptands*. Royal Society of Chemistry, ISBN 0-85186-996-3, Cambridge, UK.

Gutsche, C. D. (1998) *Calixarenes revisited: Monographs in Supramolecular Chemistry*, Royal Society of Chemistry, ISBN 978-0854045020, Cambridge, UK.

Hashimoto, T. ; Takasu, Y. ; Yamada, Y. & Ebata, T. (2006). Anomalous conformer dependent S_1 lifetime of L-phenylalanine. *Chemical Physics Letters*, Vol.421, No.1-3, pp. 227-231, ISSN 0009-2614

Hontama N.; Inokuchi Y.; Ebata T.; Dedonder-Lardeux C.; Jouvet C. & Xantheas S. S. (2010). Structure of the Calix[4]arene–(H_2O) Cluster: The World's Smallest Cup of Water. *The Journal of Physical Chemistry A*, Vol.114, No.9, pp. 2967-2972, ISSN 1089-5639

Inokuchi, Y. ; Kobayashi, Y.; Ito, T. & Ebata, T. (2007). Conformation of L-Tyrosine Studied by Fluorescence-Detected UV–UV and IR–UV Double-Resonance Spectroscopy. *The Journal of Physical Chemistry A*, Vol.111, No.17, pp. 3209-3215, ISSN 1089-5639

Inokuchi, Y.; Boyarkin, O. V.; Kusaka, R. ; Haino, T. ; Ebata, T.& Rizzo, T.R. (2011). UV and IR spectroscopic Studies of Cold Alkali Metal Ion-Crown Ether Complexes in the Gas Phase. *Journal of the American Chemical Society*, Vol.133, No.31, pp. 12256-12263, ISSN 0002-7863

Izatt, R. M.; Rytting, J. H.; Nelson, D. P.; Haymore, B. L. & Christensen, J. J. (1969). Binding of Alkali Metal Ions by Cyclic Polyethers: Significance in Ion Transport Processes. *Science*, Vol.164, No.3878, pp. 443-444, ISSN 0036-8075

Kaneko, S., Inokuchi, Y., Takayuki Ebata, T.; Aprà E. & Xantheas, S. S. (2011). Laser Spectroscopic and Theoretical Studies of Encapsulation Complexes of Calix[4]arene. *The Journal of Physical Chemistry A*, Vol.115, No.40, pp. 10846-10853, ISSN 1089-5639

Kendall, R.; Dunning, T., Jr. & Harrison, R. (1992). Electron affinities of the first-row atoms revisited. Systematic basis sets and wave functions. *Journal of Chemical Physics*, Vol.96, No.9 pp. 6796-6806, ISSN 0021-9606

Kusaka, R. ; Inokuchi, Y. & Ebata, T. (2007). Laser spectroscopic study on the conformations and the hydrated structures of benzo-18-crown-6-ether and dibenzo-18-crown-6-ether in supersonic jets. *Physical Chemistry Chemical Physics*, Vol.9, No.32, pp. 4452-4459, ISSN 1463-9076

Kusaka, R. ; Inokuchi, Y. & Ebata, T. (2008). Structure of hydrated clusters of dibenzo-18-crown-6-ether in a supersonic jet—encapsulation of water molecules in the crown cavity. *Physical Chemistry Chemical Physics*, Vol.10, No.41, pp. 6238-6244, ISSN 1463-9076

Kusaka, R. ; Inokuchi, Y. & Ebata, T. (2009). Water-mediated conformer optimization in benzo-18-crown-6-ether/water system. *Physical Chemistry Chemical Physics*, Vol.11, No.40, pp. 9132-9140, ISSN 1463-9076

Kusaka, R.; Kokubu, S.; Inokuchi, Y.; Haino, T. & Ebata, T. (2011). Structure of host-guest complexes between dibenzo-18-crown-6 and water, ammonia, methanol, and acetylene: Evidence of molecular recognition on the complexation. *Physical Chemistry Chemical Physics*, Vol.13, No.15, pp. 6827-6836, ISSN 1463-9076

Kuzmicz, R.; Dobrzycki, L.;Wozniak, K.; Benevelli, F.; Klinowski, J. & Kolodziejski, W. (2002). X-ray diffraction and ^{13}C solid-state NMR studies of the solvate of tetra(C-undecyl)calix[4]resorcinarene with dimethylacetamide. *Physical Chemistry Chemical Physics*, Vol.4, No.11, pp. 2387-2391, ISSN 1463-9076

Lee, K. T; Sung, J.; Lee, K. J. & Kim, S. K. (2002). Resonant two-photon ionization study of jet-cooled amino acid: L-phenylalanine and its monohydrated complex. *Journal of Chemical Physics*, Vol.116, No.19 pp. 8251-8254, ISSN 0021-9606

Lehn, J. -M. (1995). *Supramolecular Chemistry: Concepts and Perspectives*, Wiley-VCH, ISBN 978-3527293117, Weinheim, Germany

Lide, R. D. (2001). *CRC Handbook of Chemistry and Physics* (82nd Ed.), CRC Press, pp. 9-45, ISBN 978-0849304828

Molins, M. A.; Nieto, P. M.; Sanchez, C.; Prados, P.; Mendoza, J. De. & Pons, M. (1992). Solution structure and conformational equilibria of a symmetrical calix[6]arene. Complete sequential and cyclostereospecific assignment of the low-temperature

NMR spectra of a cycloasymmetric molecule. *The Journal of Organic Chemistry*, Vol.57, No.25, pp. 6924-6931, ISSN 0022-3263

More, M. B.; Ray, D. & Armentrout, P. D. (1999). Intrinsic Affinities of Alkali Cations for 15-Crown-5 and 18-Crown-6: Bond Dissociation Energies of Gas-Phase M$^+$–Crown Ether Complexes. *Journal of the American Chemical Society*, Vol.121, No.2, pp. 417-423, ISSN 0002-7863

Møller, C. & Plesset, M. S. (1934). Note on an Approximation Treatment for Many-Electron Systems. *Physical Review*, Vol.46, No.7 pp. 618-622, ISSN 0031-899X

Pedersen, C. J. (1967). Cyclic polyethers and their complexes with metal salts. *Journal of the American Chemical Society*, Vol.89, No.26, pp. 7017-7036, ISSN 0002-7863

Pedersen, C. J. & Frensdorff, H. K. (1972). Macrocyclic Polyethers and Their Complexes. *Angewandte Chemie International Edition*, Vol.11, No.1, pp. 16-25, ISSN 0570-0833

Peiris, D. M.; Yang, Y.; Ramanathan, R.; Williams, K. R.; Watson, C. & Eyler, J. R. (1996). Infrared multiphoton dissociation of electrosprayed crown ether complexes. *International Journal of Mass Spectrometry and Ion Processes*, Vol.157-158, pp. 365-378, ISSN 0020-7381

Purse, B.W.; Gissot, A. & Rebek, J. Jr (2005). A Deep Cavitand Provides a Structured Environment for the Menschutkin Reaction. *Journal of the American Chemical Society*, Vol.127, No.32, pp. 11222-11223, ISSN 0002-7863

Robertson, E. & Simons, J. P. (2001). Getting into shape: Conformational and supramolecular landscapes in small biomolecules and their hydrated clusters. *Physical Chemistry Chemical Physics*, Vol.3, No.1, pp. 1-18, ISSN 1463-9076

Snoek, L. C.; Robertson, E. G.; Kroemer, R. T. & Simons, J. P. (2000). Conformational landscapes in amino acids: infrared and ultraviolet ion-dip spectroscopy of phenylalanine in the gas phase. *Chemical Physics Letters*, Vol.321, No.1-2, pp. 49-56, ISSN 0009-2614

Szejtli, J (1988). *Cyclodextrin Technology: Topics in Inclusion Science*, Springer, ISBN 978-9027723147, Dordrecht, The Netherlands

Tanabe, S.; Ebata, T.; Fujii, M. & Mikami, N. (1993). OH stretching vibrations of phenol—$(H_2O)_n$ (n=1-3) complexes observed by IR-UV double-resonance spectroscopy. *Chemical Physics Letters*, Vol.215, No.4, pp. 347-352, ISSN 0009-2614

Thallapally, P. K.; Lloyd, G. O.; Atwood ,J. L. & Barbour , L.J. (2005) Diffusion of Water in a Nonporous Hydrophobic Crystal. *Angewandte Chemie International Edition*, Vol.44, No.25 , pp. 3848-3851, ISSN 0570-0833

Von Helden, G. ; Compagnon, I.; Blom, M. N.; Frankowski, M. ; Erlekam, U. ; Oomens, J.; Brauer, B. ; Gerber, R. B. & Meijer, G. (2008). Mid-IRspectra of different conformers of phenylalanine in the gas phase. *Physical Chemistry Chemical Physics*, Vol.10, No.9, pp. 1248-1256, ISSN 1463-9076

Watanabe, T. ; Ebata, T.; Tanabe S. & Mikami, N. (1996). Size-selected vibrational spectra of phenol-$(H_2O)_n$ (n=1–4) clusters observed by IR-UV double resonance and stimulated Raman-UV double resonance spectroscopies. *Journal of Chemical Physics*, Vol.105, No.2 pp. 408-419, ISSN 0021-9606

Zwier, T. S. (1996). THE SPECTROSCOPY OF SOLVATION IN HYDROGEN-BONDED
    AROMATIC CLUSTERS. *Annual Review of Physical Chemistry*, Vol.47, pp. 205-241,
    ISSN 0066-426X

3

High Pressure Raman Spectra of Amino Acid Crystals

Paulo de Tarso Cavalcante Freire,
José Alves Lima Júnior, Bruno Tavares de Oliveira Abagaro,
Gardênia de Sousa Pinheiro, José de Arimatéa Freitas e Silva,
Josué Mendes Filho and Francisco Erivan de Abreu Melo
Departamento de Física, Universidade Federal do Ceará,
Brazil

1. Introduction

Amino acids are molecules with general formula $HCCO_2^-NH_3^+R$, where R is a lateral chain characteristic of each molecule, which form the proteins of all living beings. Due the fact that both hydrogen bonding (HB) interactions play a central role on the secondary structure of proteins and the amino acids in crystal structure present complex networks of HB, they have been studied extensively in the last years (Barthes et al., 2004; Boldyreva et al., 2003a, 2003b, 2004, 2005a, 2007a, 2007b; Dawson et al., 2005; Destro et al., 1988; Façanha Filho et al., 2008, 2009; Freire, 2010; Funnel et al., 2010; Goryainov et al., 2005, 2006; Harding & Howieson, 1976; Hermínio da Silva et al., 2009; Lima et al., 2008; Migliori et al., 1988; Murli et al., 2003; Olsen et al., 2008; Sabino et al., 2009; Teixeira et al., 2000; Tumanov et al., 2010; Yamashita et al., 2007). These studies can be seen as important background to understand both static structure and dynamics properties of proteins such as denaturation, renaturation, folding itself, changes of folds, among others (Freire, 2010). The simplest protein amino acid is α-glycine, which was investigated by Raman spectroscopy under high pressure conditions, being discovered that up to 23 GPa the crystal structure does not undergo any phase transition, although modifications in the Raman spectra indicate changes in the intra-layer HB interactions (Murli et al., 2003). In this chapter we investigate the effect of high hydrostatic pressure on L-histidine hydrochloride monohydrate (HHM) and L-proline monohydrate (PM) crystals, in particular observing the effect of pressure on the vibrations related to hydrogen bonds observed in these amino acid crystals.

2. State of the art

Many works have investigated high pressure vibrational and structural properties of amino acid crystals. The simplest amino acid is glycine, which at atmospheric pressure presents at least three different polymorphs. When pressure is applied to the different polymorphs, different results are obtained. In this way, when α–glycine is submitted to high pressure, up to 23 GPa, no structural modification is verified (Murli et al., 2003). Relatively short N-H...O hydrogen bonds form layers parallel to the ac plane and the layers are connected by much weaker bifurcated N-H...O hydrogen bond forming anti-parallel double layers (Murli et al., 2003). The difficult of rearranging the double layers in the crystal structure is an explanation for the stability of α-glycine with respect to pressure (Boldyreva, 2007b). Such a fact, the stability of the structure under high pressure conditions, is very different from what is observed with β- and γ-glycine. The β-glycine has a crystal structure very similar to the α-glycine, although this last polymorph is most stable and should be obtained from the β-polymorph under humid conditions (Dawson et al., 2005). For β-glycine it was observed through both Raman and polarizing spectroscopies a reversible phase transition at 0.76 GPa (Goryainov et al., 2005). Such a phase transition is accompanied by pronounced changes in the Raman spectra of the material, in particular by jumps and kinks at the curves of frequency versus pressure (for the band associated to NH_3 rock, the jump is higher than 10 cm^{-1}). Additionally, the transition is characterized by a rapid propagation of an interphase boundary accompanied by the crack formation in the crystal as verified by authors of Ref. (Goryainov et al., 2005). On decompression, the high pressure phase (β'-glycine) transforms back to the β-glycine without hysteresis (Goryainov et al., 2005), which is not a general result among amino acid crystals.

While the α- and β-forms of glycine crystallize in a monoclinic structure, respectively, with space groups $P2_1/n$ and $P2_1$, γ-glycine crystallizes in a trigonal symmetry $(P3_1)$ (Boldyreva et al., 2003a). Under ambient conditions the α- and γ-forms of glycine are stable for a very long time, but under high temperature (T ~ 440 K) it is observed a phase transition from the γ- to the α-polymorph of glycine. On the other hand, it is interesting to note that even the α-form being thermodynamically slightly less stable than the γ-form at low temperatures, its transformation to the γ-form is apparently kinetically hindered (Boldyreva et al., 2003b). Related to the behavior of γ-glycine under hydrostatic pressure some studies have investigated this subject. Under the scrutiny of X-ray diffraction authors of Ref. (Boldyreva et al., 2004) observed that γ-glycine undergoes a phase transition beginning at 2.7 GPa characterized by an abrupt change in the unit cell volume; such a phase transition is not completed even to pressures of 7.8 GPa. On compression up to 2.5 GPa, γ-glycine structure is changing anisotropically in such a way that the a/c ratio decreases. At 2.7 GPa the reflection of a new phase, δ-glycine, begins to appear, but when pressure is released even at ambient pressure, the new phase did not disappears completely. In other words, part of the high pressure δ-glycine phase remains down to the atmospheric pressure and the γ – δ transformation was not completely reversible (Boldyreva et al., 2005a). On decompression of the γ and δ phases, additionally, it was observed the appearance of another new phase at 0.62 GPa, ζ–glycine, which could be observed both by Raman and optical spectroscopies (Goryainov et al., 2006).

Among the amino acids, L-alanine is the most studied crystalline system and although a great number of works has visited its physical and chemical properties, there are some interesting non conclusive questions related to it. For example, although there is no indication for occurrence of temperature induced phase transition, it is known that the c lattice parameter of L-alanine increases with decreasing temperature (Destro et al., 1988) by a step-wise dynamics (Barthes et al., 2004). Under low temperature, authors of Ref. (Migliori et al., 1988) have observed an unusual behavior of the intensities of the low wavenumber modes at 41 and 49 cm^{-1}, associating the phenomenon to the occurrence of localized vibrational states. Such modes have also an intriguing behavior with high pressure conditions: between 0 and 2.3 GPa the lowest wavenumber band increases intensity while the band at 49 cm^{-1} decreases; above the critical pressure of 2.3 GPa, an inverse effect is observed, e.g., the lowest wavenumber band decreases in intensity while the other band increases its intensity (Teixeira et al., 2000). Beyond this, under the scrutiny of the behavior of lattice modes of L-alanine through Raman spectroscopy it was reported the evidence of a structural phase transition at ~ 2.3 GPa (Teixeira et al., 2000), that seems to be confirmed by X-ray diffraction measurements (Olsen et al., 2008). However, very recent works reinterpreted X-ray diffraction measurements as conformational changes of the ammonia group (Funnel et al., 2010; Tumanov et al., 2010). Additionally, (i) the Raman scattering data were not correlated with a structural phase transition, only with continuous changes in the intermolecular interactions (Tumanov et al., 2010); (ii) at about 2 GPa the cell parameters a and b become accidentally equal to each other (Funnel et al., 2010; Tumanov et al., 2010), but maintaining the same orthorhombic structure ($P2_12_12_1$), differently from the work of ref. (Olsen et al., 2008) that interpreted the X-ray diffraction results as an orthorhombic → tetragonal phase transition.

There are three other aliphatic amino acids, which were investigated by Raman spectroscopy under high pressure conditions: L-leucine, L-isoleucine and L-valine. At atmospheric pressure and room temperature, L-leucine ($C_6H_{13}NO_2$) crystallizes in a monoclinic structure (C_2^2) (Harding & Howieson, 1976), although there are reports of a second phase for temperatures higher than 353 K (Façanha Filho et al., 2008) and a third polymorph resulting from residues of a solution at 200 Ma (Yamashita et al., 2007). A series of modifications on the Raman spectrum of L-leucine crystal were observed when it was submitted to high pressure conditions (Façanha Filho et al., 2009). The modifications occur in three different pressure ranges: (i) between 0 and 0.46 GPa, (ii) between 0.8 and 1.46 GPa, and (iii) at around 3.6 GPa. The first modification observed in the Raman spectra involves motions of the CH and CH$_3$ units, as can be understood from the behavior of the bands in the high wavenumber region (about 3000 cm^{-1}). It is worth to note that such a change is also associated to the hydrogen bond changes, because an increasing in the line width of a band associated with the torsion of CO_2^- is verified across the pressure of 0.46 GPa, although there is no great change in the lattice modes. Differently, the changes observed between 0.8 and 1.46 GPa occurs both in the internal modes and in the lattice modes of the crystal, indicating a structural phase transition undergone by L-leucine. Finally, around 3.6 GPa change of the slopes of the frequency $versus$ pressure plots associated with CO_2^- moieties were observed, pointing to changes associated to hydrogen bonds (Façanha Filho et al., 2009).

The Raman spectrum of L-valine ($C_5H_{11}NO_2$) under high pressure conditions, presents several changes between 0.0 and 6.9 GPa (Hermínio da Silva et al., 2009). In particular, an extraordinary increase of intensity of the C – H stretching bands is verified at about 3 GPa and a decrease of intensity is observed at ~ 5.3 GPa. Simultaneously, discontinuities are observed in the frequency $versus$ pressure plots for all modes of the Raman spectrum in these two pressure values, indicating possible phase transitions undergone by the crystal. L-isoleucine ($C_6H_{13}NO_2$) was another aliphatic amino acid whose Raman spectra were investigated under high pressure (Sabino et al., 2009). From this preliminary study, it were observed changes on bands associated with both the rocking of NH$_3^+$, r(NH$_3^+$), and the rocking of CO_2^-, r(CO_2^-), as well as to lattice modes at ~ 2.3 GPa and 5.0 GPa. Such modifications in L-isoleucine were associated with conformational change of molecules or to a phase transition undergone by the crystal. However, a confirmation of the occurrence of phase transitions for L-valine, L-leucine and L-isoleucine through X-ray diffraction or neutron diffraction is still lacking.

Two sulfur amino acids were investigated under high pressure conditions, L-methionine (Lima et al., 2008) and L-cysteine (Minkov et al., 2008, 2010; Moggach et al., 2006; Murli et al., 2006). L-methionine ($C_5H_{11}NO_2S$) crystallizes in a monoclinic structure and under compression undergoes a phase transition at about 2.2 GPa. This modification is realized by the observation of the appearance of a very strong peak between the bands associated to stretching of SC, $\nu(SC)$, and wagging vibration of CO_2^-, $\omega(CO_2^-)$. At P = 2.7 GPa, the intensity of the bands $\nu(SC)$ and $\omega(CO_2^-)$ goes to zero. Additionally, two new structures are observed around 540 cm^{-1}, in the region where it is expect to be observed the band associated to the rocking of CO_2^-, $r(CO_2^-)$. Such a picture was interpreted as a phase transition undergone by L-methionine crystal, with a hysteresis of about 0.8 GPa. On the other hand, L-cysteine can crystallize in two different polymorphs with orthorhombic and monoclinic symmetries (Minkov et al., 2010). For the monoclinic polymorph of L-cysteine it was observed phase transitions at about 2.9 and 3.9 GPa with the changes in the Raman spectra suggesting that the hydrogen bond network is distorted and the S-H...O bonding dominates over S-H...S bonding at high pressures. For the orthorhombic polymorph, a series of different phase transition occurs, but with no evidence that it transforms into the most dense monoclinic polymorph which is also stable at atmospheric pressure. It is worth to note that the phase transitions in the orthorhombic L-cysteine involve changes in molecular conformation while the pressure-induced phase transitions in the monoclinic phase are mainly related to changes in the hydrogen bond network (Minkov et al., 2010).

L-threonine ($C_4H_9NO_3$) (Silva et al., 2000) was investigated up to 4.3 GPa through Raman spectroscopy. From this study it was observed several modifications in the Raman spectrum, including the region associated to the lattice mode vibrations, mainly between 2 and 2.2 GPa. Other modifications above 3 GPa were also observed but no X-ray diffraction experiment was performed up to now, which would confirm the occurrence of eventual phase transition undergone by L-threonine crystal.

L-serine was also investigated under high pressure conditions (Boldyreva et al., 2005b, 2006a, 2006b; Moggach et al., 2005, 2006). From x-ray diffraction measurements (Moggach et al., 2005) it was observed that at ~ 4.8 GPa the crystal undergoes a phase transition with changes in the hydrogen bond network: while the low pressure phase is characterized by OH...OH hydrogen bond chains, in the high pressure phase it is observed shorter OH...carboxyl interactions. Yet, Ref. (Moggach et al., 2005) shows that the phase transition occurs with change in two torsion angles but with any major changes in the orientations of the molecules in the unit cell. Study of Ref. (Boldyreva et al., 2006b) confirms the phase transition previously reported in Ref. (Moggach et al., 2005), although with a transition pressure value of ~ 5.3 GPa, and points the existence of another phase transition at 7.8 GPa. In this second phase transition new OH...O(CO) hydrogen bond and a new NH...OH bond are formed, showing that OH-group becomes both a proton donor and a proton acceptor.

Finally, finishing this picture of the state of the art of vibrational and structural aspects of amino acid crystals under high pressure, we briefly discuss results on L-asparagine monohydrate, which was investigated both by Raman spectroscopy (Moreno et al., 1997) and by X-ray diffraction experiments (Sasaki et al., 2000). Both studies point to the occurrence of three different phase transitions between 0.0 and 1.3 GPa, which constitute, up to now, the most unstable amino acid crystal structure.

In all these investigations, the occurrence of phase transitions involving change in the dimension of the several HB in the unit cell seems to be the role. An extension of studies of amino acid crystals under high pressure conditions investigated by Raman spectroscopy for L-histidine.HCl.H$_2$O and L-proline monohydrate is given here.

3. Experimental

The samples of HHM were grown from aqueous solution by the slow evaporation method at controlled temperature. The samples of PM polycrystals were obtained from reagent grade (Sigma Aldrich) and used without further purification. Raman experiments on HHM were performed in the backscattering geometry employing a Jobin Yvon Triplemate 64000 micro-Raman system equipped with a N$_2$-cooled charge-coupled device (CCD) detection system, while for Raman experiments on PM it was employed an HR 460 Jobin Yvon spectrometer. The slits were set for a 2 cm^{-1} spectral resolution. Raman spectra were excited with the 514.5 nm line of an argon ion laser. The high-pressure experiments at room temperature were performed on a small piece of sample compressed using a diamond anvil cell (model NBS — National Bureau of Standards). For the Raman experiments on PM it was used a membrane diamond anvil cell (MDAC) with a 400 µm culet diameter diamonds as the pressure device. A 150 µm-diameter hole in a stainless-steel (200 µm of initial thickness preindented to 40 µm) was loaded with argon, while a 200 µm-diameter hole in a stainless-steel with a 230 µm of initial thickness was loaded with mineral oil in the study of L-histidine.HCl.H$_2$O. An Olympus microscope lens with a focal distance f = 20.5 mm and a numerical aperture 0.35 was used to focus the laser beam on the sample surface, which was located in the pressure cell. The pressure in the cell was monitored using the standard shifts of the Cr^{3+}:Al$_2$O$_3$ emission lines.

4. Results

4.1 L-proline monohydrate
In this sub-section we present investigation of polycrystalline PM under hydrostatic pressure up to 11.8 GPa. It is important to state that the sample used in the experiment had a small quantity of anhydrous phase (about 8%, according to the Rietveld refinement) in such a way that the structure can be considered as monoclinic L-proline monohydrate.

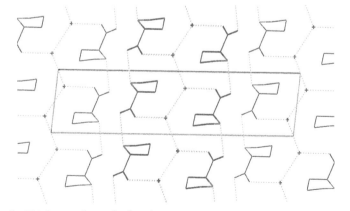

Fig. 1. Unit cell of PM seen along the b-axis.

PM (see Figure 1) crystal is found in a monoclinic structure, space group C2with a = 20.43 Å, b = 6.19 Å, c = 5.14 Å and β = 95.79° and Z=4 (Seijas et al., 2010). In such a material the hydrogen bonds play a special role in the stability of the crystal structure. X-ray diffraction study shows that there are interactions between amino and carboxylate groups through the hydrogen bonds N1-H1...O1, N1-H1...O2 and N1-H2...O1, with H1 atom acting as a bifurcated donor taking part in two hydrogen bonds. These hydrogen bonds link the proline molecules forming pairs of chains in opposite directions along the c axis. The water molecules form zigzag chains of O-H...O hydrogen bonds also parallel to the c axis (Seijas et al., 2010).

Figure 2 presents the Raman spectra of polycrystalline PM crystal in the spectral range 10 – 250 cm^{-1} for pressures from 0.0 to 11.8 GPa. The spectrum taken at 0.0 GPa shows a complex profile indicating the occurrence of several bands; in fact, fitting with Lorentzian curves

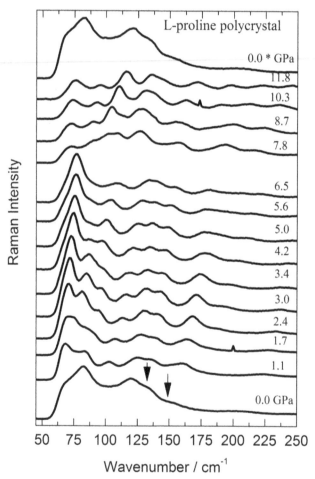

Fig. 2. Raman spectra of PM under several pressure conditions from 0.0 to 11.8 GPa in the spectral range 50 – 250 cm^{-1}.

points to the existence of six bands between 10 and 160 cm^{-1}. When pressure is increased great changes are observed in the Raman profile of the spectrum. At 1.1 GPa the Raman spectrum of L-proline is very different from that obtained at 0.0 GPa: (i) the mode peaked at ~ 68 cm^{-1} becomes the most intense band in the spectrum at 1.1 GPa; (ii) the most intense peak in the spectrum taken at 0.0 GPa decreases intensity and in the spectrum taken at 1.1 GPa it is only the second most intense peak in the region 10 – 250 cm^{-1}; (iii) the peaks marked by arrows in the spectrum at 0.0 GPa increase intensity. Such a set of modifications point to the possibility of the occurrence of a phase transition undergone by polycrystalline L-proline for pressures below 1.1 GPa. Results on other spectral regions of the Raman spectrum will reinforce this hypothesis.

It is observed a gradual change of intensity of all bands appearing in the Raman spectra of Figure 2 as well as continuous blue shifts of all wavenumbers when pressure is increased from 11.1 to 6.5 GPa. The gradual change of the profiles of the Raman spectrum can be interpreted as continuous changes of the conformation of the molecules of L-proline in the unit cell of the crystal. However, between 6.5 and 7.8 GPa it is observed a strong change in the profile of the Raman spectrum in the low wavenumber region. In particular, the most intense band observed in the spectrum taken at 6.5 GPa lost intensity becoming one of the peaks with lowest intensity. Additionally, jump in the wavenumber of almost all bands are also verified in this pressure interval (6.5 - 7.8 GPa). As a consequence, we have interpreted the changes of the Raman spectrum as a phase transition undergone by polycrystalline L-proline between 6.5 and 7.8 GPa. Other evidences will be shown in the next paragraphs. Finally, when we decrease pressure from the highest value obtained in our experiments down to atmospheric pressure we observe that the original spectrum – bottom spectrum in Figure 2 – is recovered (the final spectrum is marked as 0.0 * GPa).

Figure 3 presents the Raman spectra of PM crystal for several pressures in the spectral range 200 – 750 cm^{-1}. According to Ref. (Herlinger & Long, 1970) (see also Table 1) the band recorded close to 294 cm^{-1} in the 0.0 GPa spectrum can be associated to a deformation vibration of the skeleton of the structure, $\delta(skel)$. The bands observed at 377 and 447 cm^{-1} are associated, respectively, to the bending of CCN, $\delta(CCN)$, and rocking of CO_2^-, r(CO_2^-). In fact, for other amino acid crystals it has been reported that the r(CO_2^-) is the most intense band in the 200 – 700 cm^{-1} spectral range. However, for the other amino acids, differently from L-proline, the r(CO_2^-) mode is observed for wavenumber higher than 500 cm^{-1} (α-glycine: 503 cm^{-1} (Dawson et al., 2005); L-alanine: 532 cm^{-1}; L-valine, 542 cm^{-1} (Goryainov et al., 2005); DL-alanine, 543 cm^{-1}; etc.). Finally, in the spectrum taken at 0.0 GPa we can observe a band at 574 cm^{-1} which was associated to a bending of CO_2^-, group. Finally, bands at 641, 669 and 695 cm^{-1}, were associated, respectively, to wagging of CO_2^-, w(CO_2^-); $\delta(skel)$; and scissoring of CO_2^-, sc(CO_2^-).

By increasing the pressure we observe that between 0.0 and 1.1 GPa some important modifications are present. For example, the band $\delta(CCN)$ decreases intensity and seems to split. A clear splitting is observed for the band at 574 cm^{-1} and the r(CO_2^-) band at 447 cm^{-1} begins to present a shoulder in the high wavenumber side. Additionally, the $\delta(skel.)$ band at 669 cm^{-1} (marked by an asterisk in the spectrum of 0.0 GPa) disappears between the two lowest pressure spectra. This entire complex picture corroborates the fact that the crystal is undergoing a phase transition between 0.0 and 1.1 GPa.

If one continues to increase pressure on the PM crystal, it is possible to observe that all bands suffer blue shifts up to 6.5 GPa; also, in general terms, the linewidth increases for all bands. However, no great changes are observed, indicating that the crystal structure

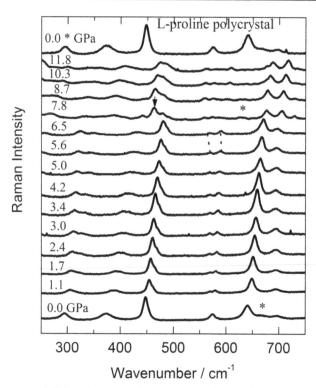

Fig. 3. Raman spectra of PM under several pressure conditions from 0.0 to 11.8 GPa in the spectral range 250 – 750 cm-1.

remains the same of that observed at 1.1 GPa. In other words, the crystal seems to be stable between 1.1 to 6.5 GPa. But when pressure arrives to 7.8 GPa great modifications are observed in the 200 – 700 cm-1 spectral region. Let us state these modifications: (i) the appearance of a band (marked by an arrow in the spectrum taken at 7.8 GPa) in the region ~ 450 cm-1; (ii) an impressive fall of intensity and a red shift of the band associated to $r(CO_2^-)$; (iii) a red shift of the two bands in the region 550 – 600 cm-1; (iv) the appearance of low intense bands (marked by an asterisk in the spectrum taken at 7.8 GPa) for wavenumbers higher than 600 cm-1; (v) the increase of intensity of the band initially assigned as scissoring of CO_2^-; (vi) the appearance of a band at ~ 725 cm-1. Again, the modifications in the spectra between 6.5 and 7.8 GPa presented in Figure 3 corroborate the fact that PM crystal undergoes a second phase transition, as discussed previously. When pressure is released down to 0.0 GPa, whose spectrum is marked by 0.0* GPa, we observe that the starting atmospheric spectrum is recovered, as already observed in Figure 2.

Figure 4 presents the Raman spectra of PM crystal under high pressure conditions, with argon as compression medium, in the region 600 to 1250 cm-1. We have cut off the peaks in the region close to 900 cm-1 because there is a peak with high intensity that makes difficult the visualization of the bands close to it; this region will be discussed in Figure 5. In the Figure 4 we observe that several peaks are well defined (the peak marked by an asterisk is due a lamp used to calibrate the spectra). It is important to state that in the spectrum at 0.0

Wavenumber (cm⁻¹)	Assignment	Wavenumber (cm⁻¹)	Assignment
68	Lattice mode	950	$v(CCN)+v(CC)$
81	Lattice mode	982	$v(CCN)+v(CC)$
99	Lattice mode	991	$v(CCN)+v(CC)$
120	Lattice mode	1031	$v(CCN)+v(CC)$
134	Lattice mode	1055	$w(CH_2)$
149	Lattice mode	1079	$w(CH_2)$
203	-	1084	$r(CH_2)$
294	δ(skel.)	1161	$t(CH_2)$
373	δ(CCN)	1172	$t(CH_2)$
447	$r(CO_2^-)$	2876	$v(CH)$
574	(CO_2^-)	2898	δ (CH)
641	$w(CO_2^-)$	2932	$v(CH)$
669	δ(skel.)	2949	$v(CH)$
695	$sc(CO_2^-)$	2971	$v(CH)$
791	δ(skel.)	2983	$v(CH)$
840	$r(CH_2)$	3006	$v(CH)$
863	$r(CH_2)$	3011	$v(CH)$
897	$r(NH_2^+)$	3041	$v(NH)$
918	$v(CCN)+v(CC)$	3068	$v(NH)$

Table 1. Wavenumber (in cm⁻¹) and approximate assignments of Raman bands for L-proline monohydrate crystal, where def., deformation; δ, bending; r, rocking ; w, wagging; sc, scissoring; v, stretching ; t, twisting.

GPa there are two peaks between 975 and 1000 cm⁻¹ and in the spectrum taken at 1.1 GPa only one band is observed. Another aspect that worth note is the fact that the band of highest energy (~ 1172 cm⁻¹ at 0.0 GPa twisting of CH_2, $t(CH_2)$) splits in the spectrum taken at 1.1 GPa. Above this last pressure the intensity of the peaks remains approximately constant up to 6.5 GPa. Between 6.5 and 7.8 GPa the Raman spectrum presents great changes. The band associated to the rocking of CH_2, $r(CH_2)$, observed initially at ~ 840 cm⁻¹, presents both a jump to high wavenumbers and an increasing in its linewidth. A splitting is observed for the band marked by an star in the spectrum recorded at 6.5 GPa and observed initially at ~ 950 cm⁻¹, which is associated to the stretching of CCN and CC units, $v(CCN)$ + $v(CC)$. If one observe the band at 1033 cm⁻¹, which is associated to the wagging of CH_2, $w(CH_2)$, interestingly, between 6.5 and 7.8 GPa it changes intensity with the band marked by a down arrow in the spectrum taken at 6.5 GPa: so, the $w(CH_2)$ band lost intensity and the band marked by a down arrow increases intensity. Similarly, the low intense bands observed between 1080 and 1200 cm⁻¹ present great changes in their intensities between 6.5 and 7.8 GPa. All these changes corroborate the modification in the crystal structure which occurs above 6.5 GPa, which are reversible when pressure is released down to 0.0 GPa.

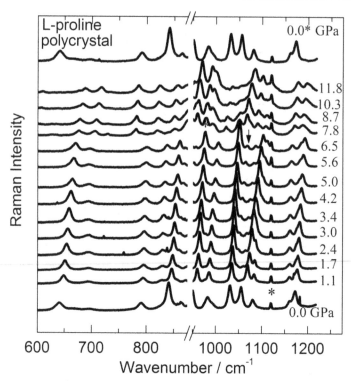

Fig. 4. Raman spectra of PM under several pressure conditions from 0.0 to 11.8 GPa in the spectral range 600 – 1250 cm-1.

Figure 5 shows the Raman spectra of PM crystal in the 800 – 1100 cm-1 spectral region. The dominant band in the spectrum taken at 0.0 GPa observed at 897 cm-1 is associated to the rocking vibration of NH_2^+, $r(NH_2^+)$. The changes occurring in the first phase transition are not impressive but we note that between 6.5 and 7.8 GPa the band lost intensity and splits into two bands. Above 7.8 GPa blue shifts of the wavenumbers are verified but no great change is present up to the highest pressure arrived in the experiment. Decreasing pressure down to 0.0 GPa again, we observe that the original spectrum is also recovered in this region.

In the spectral region presented in Figure 6, it is expected to be observed bands associated to the stretching vibrations of CH and CH_2 units. A theoretical study performed on L-methionine showed that different profiles of the Raman spectrum in this wavenumber range are associated with different conformations of the molecule. In this sense, when we observe the Raman spectra of PM as a function of pressure we note that between 0.0 and 1.1 GPa, the most intense band at 0.0 GPa split into three bands (marked by up arrows). In the spectrum taken at 1.7 GPa the separation is very clear and when pressure is further increased, the bands become clearly distinct. But, between 6.5 and 7.8 GPa, the spectrum changes abruptly, indicating that between these two pressure values the proline molecules undergo a great conformational change. Because we have observed additionally, changes in the low wavenumber region we can understand that both, conformational change and structural phase transition take place for PM crystal. As observed in the other spectral regions, the original spectrum at 0.0 GPa is recovered when pressure is released.

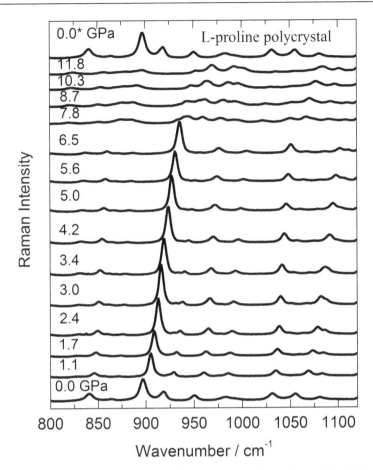

Fig. 5. Raman spectra of PM under several pressure conditions from 0.0 to 11.8 GPa in the spectral range 800 – 1100 cm⁻¹.

Data can give us some insights related to the behaviour of molecules of proline as the crystal is compressed. For example, in a previous study on α-glycine, the softening of a bending of CO_2^- was associated to a decrease in the intra-layer hydrogen bond strength, while the stiffening of the same vibration was associated to an increase in the bond strength (Dawson et al., 2005). For PM we have observed that the band associated to both bending of CO_2^-, $\delta(CO_2^-)$, and rocking of CO_2^-, $r(CO_2^-)$, increase wavenumbers from 1.1 to 6.5 GPa, and between 6.5 and 7.8 GPa decrease wavenumber; for $\delta(CO_2^-)$ the decreasing is represented by the two dashed lines in Figure 3 between the spectra of 6.5 and 7.8 GPa. We have also observed an additional increasing of wavenumbers for pressures higher than 7.8 GPa. This suggests that the intra-layer hydrogen bond strength is stiffened up to 6.5 GPa, between 6.5 and 7.8 GPa the hydrogen bond strength is softened and above 8.0 GPa, the hydrogen bond is stiffened. Additionally, our results suggest that between 6.5 and 7.8 GPa occurs a structural rearrangement in such a way that the behaviour of the hydrogen bond strengths is changed. This rearrangement is enough to change the symmetry of the crystal, because

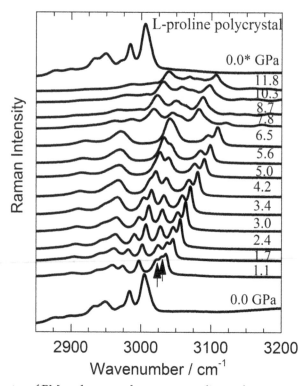

Fig. 6. Raman spectra of PM under several pressure conditions from 0.0 to 11.8 GPa in the spectral range 2850 – 3200 cm⁻¹.

associated to it we observe great modifications in the Raman spectrum of low wavenumber region (Figure 2). Yet, related to the phase transition between 6.5 and 7.8 GPa, it is impressive the increase of intensity of a band in the spectrum taken at 6.5 GPa (marked by an arrow, see Figure 4) which can be assigned as wagging of CH_2, $w(CH_2)$; it change of intensity with the band close to it at ~ 1055 cm⁻¹ at 0.0 GPa, which is also assigned as $w(CH_2)$, indicating that a conformational change in the ring may also occurs when the crystal undergoes the phase transition at 6.5 - 7.8 GPa.

It is also interesting to note that the first transition occurs from a monoclinic to a phase with lower symmetry (another monoclinic our a triclinic phase), because in several regions, including both internal and lattice mode spectral regions, the number of peaks in the spectrum taken at 1.1 GPa is greater than the number of Raman peaks in the spectrum taken at 0.0 GPa. It is also worth note that above 7.8 GPa almost all bands increases linewidth, indicating that some disorder is introduced in the high pressure phase. However, the disorder does not occur in a high degree because when pressure is released down to 0.0 GPa, the original spectrum is recovered. It is known that the amorphization process is preceded by some disorder in the crystal structure; however, due to recovered of the original Raman spectrum after releasing pressure we can infer that when we arrive to 111.8 GPa, we are far away from the pressure point where, eventually, the PM crystal should present amorphization.

4.2 L-histidine hydrochloride monohydrate

Related to the HHM (see Figure 8), which crystallizes in an orthorhombic structure belonging to the $P2_12_12_1$ (D_2^4) space group with four molecules of $C_6H_9N_3O_2.HCl.H_2O$ per unit cell, we were able to arrive up to 7.5 GPa and discovery a new polymorph of the material above 3.1 GPa. The strong bond between N-H of the imidazole ring and the carboxyl group of a neighboring molecule is the fundamental intermolecular link, resulting in a spiral arrangement along the c-axis (Donohue & Caron, 1964). The coupling of the four zwitterions in the unit cell leads to 297 optical normal modes decomposed into irreducible representations of the factor group D_2 as $\Gamma_{op} = 75\ A + 74\ (B_1 + B_2 + B_3)$ and the acoustic modes can be expressed as $\Gamma_{ac} = B_1 + B_2 + B_3$. Under high pressure conditions modifications, the Raman spectra are interpreted in terms of both conformational changes of the molecules in the unit cell and in terms of a phase transition.

Figure 9(a) presents the Raman spectra of HHM for several pressures in the spectral range from 50 to 660 cm^{-1}. The region for wavenumbers lower than 200 cm^{-1}, as previously stated, is characteristic of the lattice vibration modes. Increasing pressure we observe that the bands present blue shifts of their wavenumbers. However, between 2.7 and 3.1 GPa it is possible to note that a great change occurs in the Raman spectrum: the peak marked by a square, which is associated to torsion of CO_2^- ($\tau(CO_2^-)$), splits in two new peaks, which are marked by up arrows in the spectrum at 3.1 GPa. In Figure 9(b), which presents the plot of experimental wavenumber as function of pressure, we clearly observe the splitting of the band close to 180 cm^{-1}. Additionally we note that for other bands discontinuities of wavenumbers are observed between 2.7 to 3.1 GPa (De Sousa et al, 2011).

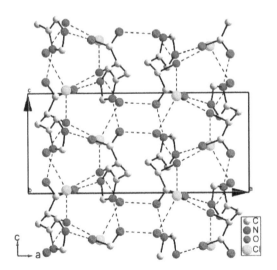

Fig. 8. Unit cell of HHM seen along the b-axis.

In this region, it is also possible to observe many internal vibrations of the histidine molecule, such as skeletal structure, at 442 and 490 cm^{-1}, torsion of NH_3^+, which is observed at 385 cm^{-1} and rocking of CO_2^- at 530 cm^{-1}. Again, as occurs with the spectra shown in the lattice mode region, modifications are observed in the spectral range 2.7 - 3.1 GPa. Among these changes we observed the disappearance of the bands at 276 and 500 cm^{-1} and the

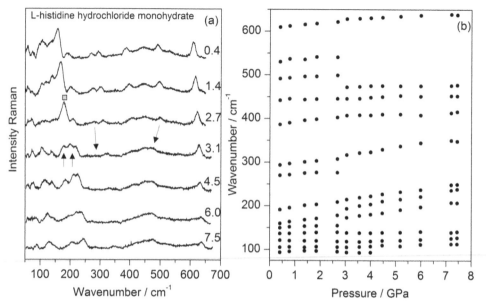

Fig. 9. (a) Raman spectra of HHM under several pressure conditions from 0.4 to 7.5 GPa in the spectral range 50 – 650 cm^{-1}; (b) wavenumber versus pressure of the bands of HHM appearing in Fig. 9(a).

appearance of a band at 473 cm^{-1} in the spectrum taken at 3.1 GPa. We additionally note that the intensity of the band associated to torsion of NH$_3^+$ decreases with increasing pressure, indicating that there occurs a conformational change of histidine molecule. Such a conformational change can be understood as consequence of the change of the intermolecular hydrogen bond length. With increasing pressure, the band initially at 385 cm^{-1}, which is associated with a torsion vibration of the NH$_3^+$ unit, shifts linearly toward higher wavenumbers; we can say that the pressure decreases the length of the hydrogen bonds, increasing the wavenumber of the band associated to torsional vibration of NH$_3^+$ unit.

Figure 10(a) presents the Raman spectra of HHM in the 650 – 1250 cm^{-1} interval for several pressure values up to 7.5 GPa. In this spectral region, it is expected to observe vibrations assigned to the deformation of CO$_2^-$, stretching of both C-C and C-N, wagging of H$_2$O and rocking of NH$_3^+$. The band observed at ~ 694 cm^{-1} is associated to a deformation vibration of CO$_2^-$, δ(CO$_2$), while the out-of-plane vibration of CO$_2^-$, γ(CO$_2$), is observed at ~ 822 cm^{-1}. The band at 807 cm^{-1} is associated with a wagging vibration of the water molecule, w(H$_2$O). The bands recorded at 1191 and 1210 cm^{-1} can be assigned as the rocking of NH$_3^+$ unit, r(NH$_3^+$). In this spectral region, some changes are observed in the range 2.7 – 3.1 GPa: (i) the disappearance of the band at 694 cm^{-1}; (ii) the appearance of a new weak band at 717 cm^{-1} (marked by a star in the Figure 10(a)); (iii) the inversion of intensity of the two bands located at 807 and 822 cm^{-1} which are related to units directly involved in the hydrogen bonds and (iv) the inversion of intensity of the two bands associated to a rocking of NH$_3^+$ unit, r(NH$_3^+$), located at 1191 and 1210 cm^{-1}. The dependence of the wavenumber versus pressure plot for this spectral region is presented in Fig. 10(b) we note that many of the observed Raman modes show a small discontinuous change in the slope in the pressure-induced variation of its wavenumber in range 2.7 – 3.1 GPa.

Figure 11(a) shows the Raman spectra of the HHM crystal in the spectral range 1700-1400 cm^{-1} for pressure from 0,0 to 7,5 GPa; Fig. 11(b) shows the respective wavenumber *vs* pressure plots. In this region, one expects to observe vibrations assigned to a deformations of the imidazole ring of the histidine molecule, asymmetric stretching of the CO_2^- units, stretching of C=O, among others. In the Figure 11(a) two modifications are observed in the range 2.7 – 3.1 GPa. The first change is the decreasing of the relative intensity of the band originally at 1433 cm^{-1} and the second one is the appearance of a band at 1642 cm^{-1}. In the wavenumber versus pressure plot presented in Figure 11(b), a subtle discontinuity occurs in the range 2.7 – 3.1 GPa, with appreciably differences in slopes, where two modes change from positive to negative slopes.

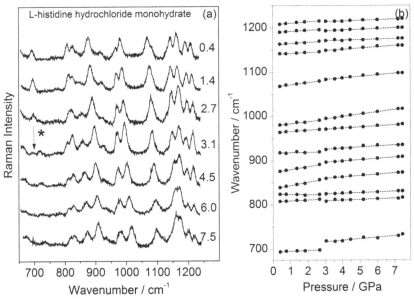

Fig. 10. (a) Raman spectra of HHM under several pressure conditions from 0.4 to 7.5 GPa in the spectral range 650 – 1250 cm^{-1}; (b) wavenumber versus pressure of the bands of HHM appearing in Fig. 10(a).

The Raman spectra of HHM crystal in the range 3450-3125 cm^{-1} are shown in the Figure 12(a) and in the Figure 12(b) the respective wavenumber *versus* pressure plots are given. In this region, we expected to observe stretching vibrations of several units of the amino acids and of the water molecule (CH, NH_3^+ and OH^-). It is important to note that for wavenumbers lower than 3000 cm^{-1} there are some modes related to the mineral oil used as compression media. In detail, the band at 3111 cm^{-1} can be associated to a stretching vibration of NH_3^+ unit, $v(NH_3^+)$, while the bands observed at 3155 and 3163 cm^{-1} are associated to a CH stretching vibration of imidazole ring. The bands observed at 3374 and 3408 cm^{-1} are related to the OH^- stretching of the water molecules. In the pressure range 2.7 – 3.1 GPa, main changes are observed such as the disappearance of the bands related to a OH^- stretching vibration (initially located at 3367 and 3393 cm^{-1}); the appearance of a band of weak intensity at 3424 cm^{-1} and discontinuities of the wavenumber curve of bands at 3155 and 3163 cm^{-1}, which are associated with the CH stretching vibration.

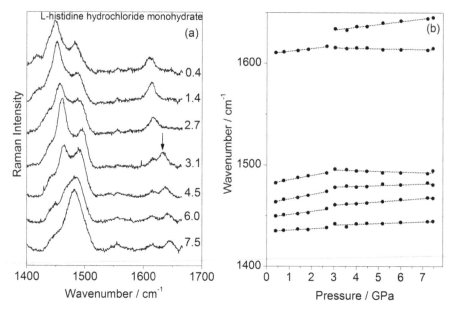

Fig. 11. (a) Raman spectra of HHM under several pressure conditions from 0.4 to 7.5 GPa in the spectral range 1400 – 1670 cm^{-1}; (b) wavenumber versus pressure of the bands of HHM appearing in Fig. 11(a).

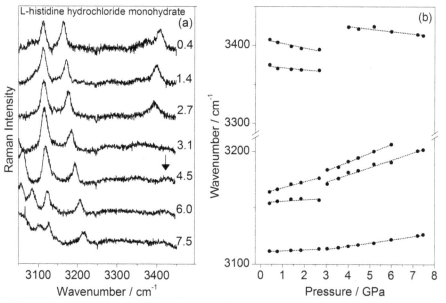

Fig. 12. (a) Raman spectra of HHM under several pressure conditions from 0.4 to 7.5 GPa in the spectral range 2950 – 3450 cm^{-1}; (b) wavenumber versus pressure of the bands of HHM appearing in Fig. 12(a).

Table 2 gives a quantitative analysis of the evolution of the Raman spectra of HHM crystal with pressure where fitting of all bands to a linear expression: $\omega = \omega_0 + \alpha \cdot P$ is furnished. In Table 2, the first column (ω_{obs}) represents the wavenumber of the Raman bands observed at room pressure, the second and third columns correspond to adjustment of data for pressure values between 0 and 2.7 GPa and the two last columns correspond to fitting of the high-pressure phase, that observed between 3.1 and 7.5 GPa.

A resume of the modifications observed in all spectral regions of the Raman spectra of HHM is as follows: (i) splitting of the band in the low-wavenumber lattice range; (ii) change in the number of bands associated to internal modes; (iii) wavenumber shifts with discontinuities; (iv) changes in the relative intensities of the vibrational bands. It is important to state that changes were observed for the deformation vibrations of the OH⁻ of the water molecule, as well as of the NH_3^+ and CO_2^- groups of the amino acid molecule. This picture indicates large conformational changes of the molecules in the unit cell.

Related to HHM, beyond the hydrogen bonds originated from the histidine molecule, there are two hydrogen bonds from the water molecule. It is interesting to note from Figure 12(b) that the wavenumber of OH⁻ stretching, $\nu(OH^-)$, decreases with increasing pressure. However, the OH units participate of hydrogen bonds, one to a carboxyl oxygen and another to the chloride ion. This means that between 0.0 and 3.1 GPa the hydrogen bond strength increases and the wavenumber of $\nu(OH^-)$ decreases in this pressure range. Also, after the phase transition, the wavenumber of $\nu(OH^-)$ suffers a jump and continues to decrease up to the highest pressure obtained in our experiments. Such a picture indicates a further increase of the hydrogen bonds formed by the water molecules and confirms that these bonds are playing an important role in the mechanism of the transition. It is interesting to note that similar changes in the low-wavenumber region were also observed for the same crystal, when subjected to low temperatures.

ω_{obs} (cm⁻¹)	Phase 1 ($0 \leq P \leq 2.7$ GPa)		Phase 2 ($3.1 \leq P \leq 7.5$ GPa)	
	ω_0 (cm⁻¹)	α (cm⁻¹)/(GPa)⁻¹	ω_0 (cm⁻¹)	α (cm⁻¹)/(GPa)⁻¹
94	94.5	0.669	93.7	-0.103
105	105.2	0.276	100.2	1.771
121	121.4	-1.087	110.9	2.092
136	135.7	1.616	136.0	0.623
149	149.7	2.128	-	-
159	156.8	8.033	148.3	8.105
-	-	-	163.9	9.960
188	189.1	7.661	190.2	8.059
270	269.7	3.093	-	-
293	291.9	6.118	294.5	7.683
385	384.6	7.236	401.5	1.675
442	442.3	1.223	442.3	1.461

ω_{obs} (cm^{-1})	Phase 1 (0 ≤ P ≤ 2.7 GPa)		Phase 2 (3.1 ≤ P ≤ 7.5 GPa)	
	ω_0 (cm^{-1})	α (cm^{-1})/(GPa)$^{-1}$	ω_0 (cm^{-1})	α (cm^{-1})/(GPa)$^{-1}$
-	-	-	470.7	0.785
491	490. 5	3.319	-	-
530	530.4	4.335	-	-
609	607.8	5.312	620.8	2.463
693	693.5	1.529	-	-
-	-	-	706.4	3.449
807	806.7	2.042	804.7	1.467
824	824.0	-0.498	820.4	1.442
838	836.1	7.335	850.4	3.124
875	873.4	6.438	888.1	2.777
917	917.3	0.810	918.7	2.356
963	963.0	2.757	963.6	2.411
980	978.7	4.489	978.7	5.206
1068	1066.7	4.790	1073.2	3.418
1141	1140.3	1.890	1140.3	2.744
1162	1161.9	2.611	1167.4	1.120
1190	1190.0	2.277	1188.5	1.637
1208	1207.9	2.859	1210.1	1.341
1434	1434.5	1.316	1438.2	0.846
1448	1448.1	3.059	1455.4	1.677
1463	1462.1	4.415	1476.0	0.778
1481	1481.3	4.238	1497.0	-0.619
1609	1609.3	2.739	1616.6	-0.374
-	-	-	1624.9	2.736
3111	3111.0	0.919	3104.5	2.914
3154	3154.6	1.287	3153.2	6.564
3164	3162.6	5.305	3159.3	7.905
3375	3373.9	-2.638	-	-
3407	3408.1	-5.745	-	-
-	-	-	3437.1	-3.169

Table 2. Wavenumber values of the bands appearing in the HHM spectrum at 0.0 GPa (ω_{obs}) and adjustment data by linear fitting ($\omega = \omega_0 + \alpha.P$) for the two phases.

5. Conclusions

The results suggested that L-proline monohydrate undergoes two phase transitions, one between 0.0 and 1.1 GPa and a second between 6.5 and 7.8 GPa. In both transitions it was possible to observed changes in the spectrum profile, discontinuities in the wavenumber versus pressure plots in the lattice region of the spectrum and the appearance of news bands. By releasing the pressure after attaining a maximum of 11.8 GPa, the atmospheric pressure spectrum was fully recovered, showing that the pressure-induced transitions undergone by polycrystalline L-proline crystal are reversible. In the first transition it was observed the splitting of bands associated to $\delta(CCN)$, $\delta(CO_2^-)$ and $\nu(CH)$ vibrations. The band associated to $\delta(skel.)$ has disappeared. In the second one, we have observed the red shift of $r(CO_2^-)$, significant changes in the intensity of the bands associated to the $sc(CO_2^-)$ and $w(CH_2)$ vibrations and the splits of $\nu(CCN) + \nu(CC)$ and $r(NH_2^+)$ that also has lost intensity. These modifications were interpreted as conformational changes of molecules in the unit cell. Because these changes involve moieties that participate in hydrogen bonds, we suppose that these hydrogen bonds have also been modified. Modifications of hydrogen bonds can trigger some molecular rearrangements that lead to structural phase transitions. In our work this hypothesis was supported by the changes that were observed in the region of the lattice modes for both phase transitions. Therefore, the results suggested two structural phase transitions, one between 0.0 and 1.1 GPa, and another between 6.5 and 7.8 GPa, accompanied by significant conformational changes of molecules in the L-proline monohydrated crystal.

The results also suggest that HHM crystal undergoes a phase transition between 2.7 and 3.1 GPa when it is submitted to high pressure conditions, up to 7.5 GPa. The main changes observed in the Raman spectra was the splitting of the band in the low-wavenumber lattice range, change in the number of bands associated to internal modes and wavenumber shifts with discontinuities.

Having described the behavior of two amino acid crystals under pressure and looked back at the past fourteen years of research we can ask to ourselves what does the future hold. Based on the studies performed we believe that the use of several techniques such as neutron and Raman scattering as well as thermal and infrared analyses can help us to give a complete picture about hydrogen bond and the behavior of amino acids under pressure. Maybe, the study of peptide under similar external conditions should be the next road to be walked.

6. Acknowledgment

Authors acknowledge CAPES, CNPq and FUNCAP for financial support.

7. References

Barthes, M.; Bordallo, H.N.; Dénoyer, F.; Lorenzo, J.-E.; Zaccaro, J.; Robert, A. & Zontone, F. (2004). Micro-transitions or breathers in L-alanine? *The European Physical Journal B*, Vol. 37, No. 3, (February 2004), pp. 375-382, ISSN 1434-6028

Boldyreva, E.V.; Drebushchak, V.A.; Drebushchak, T.N.; Paukov, I.E.; Kovalevskaya, Y.A. & Shutova, E.S. (2003a). Polymorphism of glycine, Part I. *Journal of Thermal Analysis and Calorimetry*, Vol. 73, No. 2, (August 2003), pp. 409-418, ISSN 1388-6150

Boldyreva, E.V.; Drebushchak, V.A.; Drebushchak, T.N.; Paukov, I.E.; Kovalevskaya, Y.A. & Shutova, E.S. (2003b). Polymorphism of glycine, Part II. *Journal of Thermal Analysis and Calorimetry*, Vol. 73, No. 2, (August 2003), pp. 419-428, ISSN 1388-6150

Boldyreva, E.V.; Ivashevskaya, S.N.; Sowa, H.; Ahsbahs, H. & Weber, H.-P. (2004). Effect of High Pressure on Crystalline Glycine: A New High-Pressure Polymorph. *Doklady Physical Chemistry*, Vol. 396, No. 1, (May 2004), pp. 111-114, ISSN 0012-5016

Boldyreva, E.V.; Ivashevskaya, S.N.; Sowa, H.; Ahsbahs, H. & Weber, H.-P. (2005a). Effect of hydrostatic pressure on the γ-polymorph of glycine 1. A polymorphic transition into a new δ-form. *Zeitschrift für Kristallographie, Vol.* 220, No. 1, (January 2005), pp. 50- 57, ISSN 0044-2968

Boldyreva, E.V.; Kolesnik, E.N.; Drebushchak, T.N.; Ahsbahs, H.; Beukes, J.A. & Weber, H.-P. (2005b). A comparative study of the anisotropy of a lattice strain induced in the crystals of L-serine by cooling down to 100 K or by increasing pressure up to 4.4 GPa. *Zeitschrift für Kristallographie, Vol.* 220, No. 1, (January 2005), pp. 58- 65, ISSN 0044-2968

Boldyreva, E.V.; Kolesnik, E.N.; Drebushchak, T.N.; Ahsbahs, H. & Seryotkin, Y.V. (2006a). A comparative study of the anisotropy of a lattice strain induced in the crystals of DL-serine by cooling down to 100 K, or by increasing pressure up to 8.6 GPa. A comparison with L-serine. *Zeitschrift für Kristallographie, Vol.* 221, No. 2, (2006), pp. 150- 161, ISSN 0044-2968

Boldyreva, E.V.; Sowa, H.; Seryotkin, Y.V.; Drebushchak, T.N.; Ahsbahs, H.; Chernyshev, V. & Dmitriev, V. (2006b). Pressure-induced phase transitions in crystalline L-serine studied by single crystal and gigh-resolution powder X-ray diffraction. *Chemical Physics Letters, Vol.* 429, No. 4-6, (October 2006), pp. 474-478, ISSN 0009-2614

Boldyreva, E. (2007a). Cristalline Amino Acids : A Link between Chemistry, Materials Science and Biology, In: *Models, Mysteries, and Magic of Molecules*, Boeyens, J.C.A. & Ogilvie, J.F., pp. 167-192, Springer, ISBN 978-1-4020-5940-7, Netherlands

Boldyreva, E. (2007b). High-Pressure Polymorphs of Molecular Solids: When Are They Formed, and When Are They Not? Some Examples of the Role of Kinetic Control. *Crystal Growth & Design*, Vol. 7, No. 9, (August 2007), pp. 1662-1668, ISSN 1528-7483

Dawson, A.; Allan, D.R.; Belmonte, S.A.; Clark, S.J.; David, W.I.F.; McGregor, P.A.; Parsons, S.; Pulham, C.R. & Sawyer, L. (2005). Effect of high pressure on the crystal structures of polymorphs of glycine. *Crystal Growth & Design*, Vol. 5, No. 4, (May 2005), pp. 1415-1427, ISSN 1528-7483

Destro, R.; Marsh, R.E. & Bianchi, R. (1988). A low-temperature (23K) study of L-alanine. *The Journal of Physical Chemistry*, Vol. 92, No. 4, (February 1988), pp. 966-973, ISSN 0022-3654

De Sousa, G.P.; Freire, P.T.C.; Lima Jr., J.A.; Mendes Filho, J. & Melo, F.E.A. (2011). High-pressure Raman spectra of L-histidine hydrochloride monohydrate cristal. *Vibrational Spectroscopy*, Vol. 57, No. 1, (September 2011), pp. 102-107, ISSN 0924-2031

Donohue, J. & Caron, A. (1964). Refinement of cristal structure of histidine hydrochloride monohydrate. *Acta Crystallographica*, Vol. 17, No. 9, (September 1964), pp. 1178-&, ISSN 0108-7673

Façanha Filho, P.F; Freire, P.T.C.; Lima, K.C.V.; Mendes Filho, J.; Melo, F.E.A. & Pizani, P.S. (2008). High temperature Raman spectra of L-leucine crystals. *Brazilian Journal of Physics*, Vol. 38, No. 1, (March 2008), pp. 131-137, ISSN 0103-9733

Façanha Filho, P.F.; Freire, P.T.C.; Melo, F.E.A; Lemos, V.; Mendes Filho, J.; Pizani, P.S. & Rossatto, D.Z. (2009). Pressure-induced phase transitions in L-leucine crystal. *Journal of Raman Spectroscopy*, Vol. 40, No. 1, (January 2009), pp. 46-51, ISSN 1097-4555

Freire, P.T.C. (2010). Pressure-Induced Phase Transitions in Crystalline Amino Acids, In: *High Pressure Crystallography – From Fundamental Phenomena to Technological Applications*, Boldyreva, E. & Dera, P., pp. 559-572, Springer, ISBN 978-904-8192-60-1, New York, USA

Funnel, N.P.; Dawson, A.; Francis, D.; Lennie, D.R.; Marshall, W.G.; Moggach, S.A.; Warren, J.E. & Parsons, S. (2010). The effect of pressure on the crystal structure of l-alanine. *CrystEngComm*, Vol. 12, No. 9, (September 2010), pp. 2573-2583, ISSN 1466-8033

Goryainov, S.V.; Kolesnik, E.N. & Boldyreva, E.V. (2005). A reversible pressure-induced phase transition in β-glycine at 0.76 GPa. *Physica B: Condensed Matter*, Vol. 357, No. 3-4, (March 2005), pp. 340-347, ISSN 0921-4526

Goryainov, S.V.; Kolesnik, E.N. & Boldyreva, E.V. (2006). Raman observation of a new (ζ) polymorph of glycine? *Chemical Physics Letters*, Vol. 419, No. 4-6, (February 2006), pp. 496-500, ISSN 0009-2614

Harding, M.M & Howieson, R.M. (1976). L-Leucine. *Acta Crystallographica Section B*, Vol. 32, No. 2, (February 1976), pp. 633-634, ISSN 1600-5740

Herlinger, A.W. & Long, T.V. (1970). Laser Raman and infrared spectra of amino acids and their metal complexes. 3. Proline and bisprolinato complexes. *Journal of the American Chemical Society*, Vol. 92, No. 22, (November 1970), pp. 6481-6486, ISSN 0002-7863

Hermínio da Silva, J.; Lemos, V.; Freire, P.T.C.; Melo, F.E.A.; Mendes Filho, J.; Lima Jr., J.A & Pizani, P.S. (2009). Stability of the crystal structure of L-valine under high pressure. *Physica Status Solidi B*, Vol. 246, No. 3, (March 2009), pp. 553-557, ISSN 0370-1972

Lima, J.A.; Freire, P.T.C.; Melo, F.E.A.; Lemos, V.; Mendes Filho, J. & Pizani, P.S. (2008). High pressure Raman spectra of L-methionine Crystal. *Journal of Raman Spectroscopy*, Vol. 39, No. 10, (October 2008), pp. 1356-1363, ISSN 1097-4555

Migliori, A.; Maxton, P.M.; Clogston, A.M.; Zirngiebl, E. & Lowe, M. (1988). Anomalous temperature dependence in the Raman spectra of l-alanine: Evidence for dynamic localization. *Physical Review B*, Vol. 38, No. 18, (December 1988), pp. 13464-13467, ISSN 1098-0121

Minkov, V.S.; Krylov, A.S.; Boldyreva, E.V.; Goryainov, S.N.; Bizyaev, S.N. & Vtyurin, A.N. (2008). Pressure-Induced Phase Transitions in Crystalline L- and DL-Cysteine. *The Journal of Physical Chemistry B*, Vol. 112, No. 30, (July 2008), pp. 8851-8854, ISSN 1520-6106

Minkov, V.S.; Goryainov, S.V. ;Boldyreva, E.V. & Görbitz, C.H (2010). Raman study of pressure-induced phase transitions in the crystals of orthorhombic and monoclinic polymorphs of L-cysteine: dynamics of the side-chain. *Journal of Raman Spectroscopy*, Vol. 41, No. 12, (December 2010), pp. 1458-1468, ISSN 1097-4555

Moggach, S.A.; Allan, D.R.; Morrison, P.A.; Parsons, S. & Sawyer, L. (2005). Effect of pressure on the crystal structure of L-serine-I and the crystal structure of L-serine-II at 5.4 GPa. *Acta Crystallographica Section B*, Vol. 61, No. 1, (February 2005), pp. 58-68, ISSN 0108-7681

Moggach, S.A.; Marshall, W.G. & Parsons, S. (2006). High-pressure neutron diffraction study of L-serine-I and L-serine-II, and the structure of L-serine-III at 8.1 GPa. *Acta Crystallographica Section B*, Vol. 62, No. 5, (October 2006), pp. 815-825, ISSN 0108-7681

Moreno, A.J.D.; Freire, P.T.C.; Melo, F.E.A.; Araújo-Silva, M.A.; Guedes, I.; Mendes Filho, J. (1997). Pressure-induced phase transitions in monohydrated l-asparagine aminoacid crystals. *Solid State Communications*, Vol. 103, No. 12, (September 1997), pp. 655-658, ISSN 0038-1098

Murli, C.; Sharma, S.K.; Karmakar, S. & Sikka, S.K. (2003). α-Glycine under high pressures: a Raman scattering study. *Physica B: Condensed Matter*, Vol. 339, No. 1, (November 2003), pp. 23-30, ISSN 0921-4526

Murli, C.; Vasanthi, R. & Sharma, S.M. (2006). Raman spectroscopic investigations of DL-serine and DL-valine under pressure. *Chemical Physics*, Vol. 331, No. 1, (December 2006), pp. 77-84, ISSN 0301-0104

Olsen, J.S.; Gerward, L.; Freire, P.T.C.; Mendes Filho, J.; Melo, F.E.A. & Souza Filho, A.G. (2008). Pressure-induced phase transitions in L-alanine crystals. *Journal of Physics and Chemistry of Solids*, Vol. 69, No. 7, (July 2008), pp. 1641-1645, ISSN 0022-3697

Sabino, A.S.; De Sousa, G.P.; Luz-Lima, C.; Freire, P.T.C.; Melo, F.E.A. & Mendes Filho, J. (2009). High-pressure Raman spectra of L-isoleucine crystals. *Solid State Communications*, Vol. 149, No. 37-38, (October 2009), pp. 1553-1556, ISSN 0038-1098

Sasaki, J.M.; Freire, P.T.C.; Moreno, A.J.D.; Melo, F.E.A.; Guedes, I.; Mendes Filho, J.; Shu, J.; Hu, J. & Mão, H.K. (2000). Single crystal X-ray diffraction in monohydrate L-asparagine under hydrostatic pressure. *Science and Technology of High Pressure, Proceedings of AIRAPT-17*. Hyderabad, India, 2000

Seijas, L.E.; Delgado, G.E.; Mora, A.J.; Fitch, A.N. & Brunelli, M. (2010). On the Crystal structures and hydrogen bond patterns in proline pseudopolymorphs. *Powder Diffraction*, Vol. 25, No. 3, (September 2010), pp. 235-240, ISSN 0885-7156

Silva, B.L.; Freire, P.T.C.; Melo, F.E.A.; Mendes Filho, J.; Pimenta, M.A. & Dantas, M.S.S. (2000). High-pressure Raman spectra of L-threonine crystal. *Journal of Raman Spectroscopy*, Vol. 31, No. 6, (June 2000), pp. 519-522, ISSN 1097-4555

Teixeira, A.M.R.; Freire, P.T.C.; Moreno, A.J.D.; Sasaki, J.M.; Ayala, A.P.; Mendes Filho, J. & Melo, F.E.A. (2000). High-pressure Raman study of l-alanine crystal. *Solid State Communications*, Vol. 116, No. 7, (October 2000), pp. 405-409, ISSN 0038-1098

Tumanov, N.A.; Boldyreva, E.V.; Kolesov, B.A.; Kurnosov, A.V. & Quesada Cabrera, R. (2010). Pressure-induced phase transitions in L-alanine, revisited. *Acta Crystallographica Section B*, Vol. 66, No. 4, (August 2010), pp. 458-371, ISSN 1600-5740

Yamashita, M.; Inomata, S.; Ishikawa, K.; Kashiwagi, T.; Matsuo, H.; Sawamura, S. & Kato, M. (2007). A high-pressure polymorph of L-α-leucine. *Acta Crystallographica Section E*, Vol. 63, No. 5, (May 2007), pp. o2762-o2764, ISSN 1600-5368

4

Structure Characterization of Materials by Association of the Raman Spectra and X-Ray Diffraction Data

Luciano H. Chagas[1], Márcia C. De Souza[1], Weberton R. Do Carmo[1],
Heitor A. De Abreu[2] and Renata Diniz[1]
[1]Departamento de Química, Universidade Federal de Juiz de Fora,
[2]Departamento de Química, Universidade Federal de Minas Gerais,
Brazil

1. Introduction

Supramolecular chemistry, which can be described as the chemistry beyond the covalent bonds, has become a very interesting focus of investigation in the last years. The main purpose of this kind of studies is the strategic construction of specific arrangements, and the complete comprehension of the connection between structure and physical–chemical properties. However, in the solid state there are several weak interactions, as hydrogen bonds and π-stacking, which can play a decisive role in orientation of the crystallization processes (Carlucci et al., 2003). Essentially, common features of all of the supramolecular systems are non-covalent interactions, which provide the clips linking the building blocks, leading to well organized superstructures (Yan et al., 2007). The term "Supramolecular chemistry" was introduced by Jean-Marie Lehn and it is defined as "the chemistry beyond the molecule". While a covalent bond normally has a homolytic bond dissociation energy that ranges between 20 and 100 kcal mol[-1], noncovalent interactions are generally weak and vary from less than 1 kcal mol[-1] for van der Waals forces, through approximately 25 kcal mol[-1] for hydrogen bonds, to 60 kcal mol[-1] for Coulomb interactions (Hoeben et al., 2005).

In this sense, the most important interactions in solid organic compounds are the hydrogen bonds. The kind and strength of these interactions added to the molecular arrangement are responsible for the crystal structure of these compounds. The change in production and/or storage conditions of solids may modify the hydrogen bonding design and the balance between them and van der Waals interactions. These modifications give rise several changes in the solid state, like phase transitions (Wang et al., 2009). Weak attractive forces are important in deciding the conformation of organic compounds and 3D structure of biomacromolecules. Among molecular interactions, the van der Waals force, electrostatic interactions and hydrogen bonds are the most important (Takahashi et al., 2010).

Apart of hydrogen bonds, another type of interaction that plays an important role in the design of supramolecular materials is π-π interactions. Attractive non-bonded interactions between aromatic rings are seen in many areas of chemistry, and hence are of interest to all realms of chemistry. The strength as well as the causes of these interactions, however, varies. In water the stacking interaction between aromatic molecules is mainly caused by the hydrophobic effects. Thus, π-π stacking is an important supramolecular force in molecular architecture and recognition. This force is engendered by aromatic–aromatic stacking, which enhances the stability of the complexes both in solution and in solid state. In particular, the π-π stacking interactions in solid state are widely observed in the construction of multi-dimensional structures in offset face-to-face and edge-to-face fashions; combination with coordination bonds, such π stacking interactions can be employed to build up interesting coordination supramolecular architectures (Ye et al., 2005).

The most common effect of modifications in intermolecular interactions in pharmaceutics solids is the polymorphism. Polymorphism is the capacity of the compound to crystalize in two or more different forms. This effect is very important in pharmaceutical science due to the fact that not always all polymorph has the same biodisponibility. These systems will contain drugs which are not in their thermodynamically stable state, and thus need to be stabilized against physical as well as chemical degradation (Aaltonen et al., 2008). Despite of the physical-chemical properties, the investigation of solids in pharmaceutics sciences is important because the solid formulations are the most important pharmaceutical dosage forms, due to the convenience and acceptability of solid oral forms (Aaltonen et al., 2008).

The mainly technic to identify and to investigate these interactions is the X-ray diffraction. However, vibrational spectroscopy can also be used in this kind of characterization, in special in association with crystal data. Raman and IR spectroscopy probe the solid state predominantly on the intramolecular level, although X-ray diffraction predominantly probes the "lattice level", i.e. the intermolecular level (Aaltonen et al., 2008). In addition, the computational study can contribute to the molecular or atomic level understanding the structural and thermodynamical features involved in the process of molecular recognition and supramolecular organization. The combination of these technics provides a complete description of the solid state, in special for crystalline materials. Collaborative studies by both computational and experimental groups are highly encouraged in the supramolecular chemistry (Yan et al., 2007).

2. Hydrogen bonds and vibrational spectroscopy

2.1 Description of hydrogen bonds

The hydrogen bond (or H-bond) is of great importance in natural sciences. This is related particularly to biological aspects, such as molecular recognition that could be a basis for the creation of life. The H-bond is an intermediate range intermolecular interaction between electron-deficient hydrogen and a region of high electron density. Its fundamental role in the structure of DNA and the secondary and tertiary structure of proteins is well known (Kollman and Allen, 1972). In many crystal lattices of organic compounds the H-bonds are a decisive factor governing the molecular packing. In designing of new interesting crystal structures, which is the subject of fast developing crystal engineering, one of the main parameters is the engagement of H-bonds (Sobczyk et al., 2005).

The IUPAC definition is "the hydrogen bond is a form of association between an electronegative atom and a hydrogen atom attached to a second, relatively electronegative atom. Its best considered as an electrostatic interaction heightened by a small size of hydrogen, which permits proximity of interacting dipoles or charges. Both electronegative atoms are usually (but not necessary) from the first row of the Periodic Table, i.e., N, O, or F. With a few exceptions, usually involving fluorine, the associated energies are less than 20-25 kcal mol^{-1}. Hydrogen bonds may be inter or intramolecular". This definition is limited to an already classical conception of this specific molecular interaction. It does not embrace cases such as a π-electron systems as proton acceptors, charge-assisted H-bonds of the OHO⁺ and OHO⁻ type. Moreover, in cases of strong H-bonds, a covalent nature of interaction is revealed (Sobczyk et al., 2005).

According to the simple valence bond theory, a hydrogen atom should be capable of forming only one chemical bond. In any cases, however, hydrogen is formally two-valent. According to Pauling`s definition "under certain conditions an atom of hydrogen is attracted by rather strong forces to two atoms, instead of only one, so that it may be considered to be acting as a bond between them. This additional bond is called hydrogen bond. Pauling also states that the hydrogen bond "is formed only between the most electronegative atoms". In this sense, the terms typical hydrogen bond or conventional hydrogen bond are related to the Pauling definition. There is a variety of typical H-bonds, for example, O-H ··· O existing for water dimer and formic acid dimer. The following model is useful to illustrate this definition:

$$R_1 - A^{\delta-} - H^{\delta+} \cdots B - R_2$$

where A and B are more electronegative atoms than the hydrogen atom. Most often A and B contain $2p_z$ type electrons, and hence, they may be conjugated with R_1 and R_2, if they are π-electron systems.

The atoms in the Periodic Table with electronegativity greater than that of hydrogen are C, N, O, F, P, S, Cl, Se, Br, and I; and hydrogen bonds involving all of these elements are known. On the other hand "π" hydrogen bonds involve an interaction between particularly positive hydrogen and the electrons in a double and triple bond. Atoms with electronegativity greater than hydrogen have the capability of forming A-H ··· B hydrogen bonds if B has an unshared pair of electrons, but in some cases the interaction is so weak that most chemists consider that there is no hydrogen bond formed. However, in the second half of the last century, evidence has gradually accumulated that hydrogen bonds other than the conventional hydrogen bond are ubiquitous. These include CH ··· n hydrogen bonds (n is lone pair electrons, as contrasted to π; CH ··· O, CH ··· N, etc., 2-4 kcal mol⁻¹) and XH ··· π hydrogen bonds (X = O, N, etc., 2-4 kcal mol⁻¹).

In H-bonds, the same forces are manifested as in other molecular interactions. Those forces are electrostatic in nature. However, one can distinguish a few specific features that make it possible to discriminate H-bond complex from the universal van der Waals associates and electron-donor-acceptor, named usually charge transfer complex. The most applicable is a geometric criterion. When the H-bond is compared with the van der Waals interaction, the equilibrium distance between H and B atoms is dramatically different. In the case of van der Waals interactions, the distance between H and B is close to the sum of van der Waals radii, whereas in the most frequently observed O-H ··· O bridges range between 1.6 and 2.2 Å, while from summing the H (~1.2 Å) and O (~ 1.52 Å) van der Waals radii, one obtains for $R_{H \cdots O}$ approximately 2.7 Å. Thus, a great shortening is observed.

2.2 Hydrogen bonds in carboxylic compounds

The specific interaction via H-bond is manifested in several physical properties of systems. Infrared and Raman spectroscopic evidence for hydrogen bonding is the shift of the A–H stretch in a molecule toward lower frequencies. Accordingly, infrared spectroscopy has been very important in hydrogen bonding investigations. Spectra of a hydrogen bond system present broader, more intense and shifted bands than non-hydrogen bonded systems (Emsley, 1980). The broad envelope bands region could be used to classify hydrogen bonds. In infrared spectra, broad envelope bands in 1800–900 cm^{-1} region are an evidence of short hydrogen bonds (O $\cdots$ O distance of 2.4–2.5 Å). In compounds with medium (O $\cdots$ O distance of 2.5–2.8 Å) and weak (O $\cdots$ O distance of 2.8–3.0 Å) hydrogen bonds, these broad envelopes appear in 3500–2000 cm^{-1} region (Gonzalezsanchez, 1958). The phenomena observed in IR absorption spectra related to stretching v(AH), γ(AH), and δ(AH) bending vibrations are characteristic for this kind of interaction.

Undoubtedly, these phenomena are due to the changes in charge distribution, which may be related to electron delocalization. Because both heavy-atoms components usually contain orbital able to conjugate with π-electrons systems (np orbitals or similar ones) the IR spectroscopy gives some indirect information about π-electron delocalization. The behavior of H-bonded systems in IR spectra is explained through a decrease of the force constant of v(AH) vibrations, which leads to stronger coupling with low frequency modes and coupling of Fermi resonance type, an increase of the polarizability of the H-bond. A decrease of the force constant of v(AH) vibrations and an increase of the force constants of δ(AH) and γ(AH) vibrations are caused by polarization of the AH bond and a shift of the proton toward the proton acceptor. For supramolecular polymeric structures formed by O-H $\cdots$ O hydrogen bonds, the most important spectral region is that related to medium and weak HB (Chagas et al., 2008). In this kind of structures the bands in this region are more broad and enveloped than solids with discrete HB. Figure 1 shows the infrared spectra of some compounds that present medium and weak extended HB, when the envelope bands are observed at 3500–2000 cm^{-1} region (Chagas et al., 2008; Diniz et al., 2002).

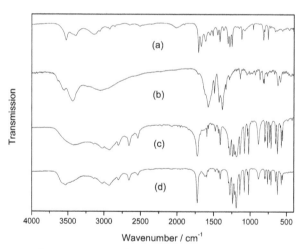

Fig. 1. Infrared spectra of compounds with extended HB: a) 1,2,4,5-benzenetetracarboxylic acid, b) triaquacopper(II) 1,2,4,5-benzenetetracarboxylate tetrahydrate, c) o-sulfobenzoic acid and d) hexaaquazinc(II) hydrogen o-sulfobenzoate.

Infrared spectra of hydrate compounds generally show some broad and intense envelope bands in the range of hydrogen-bond (HB) modes. In contrast, the water modes are very weak in Raman spectra. As a result of this, the Raman spectra of hydrate compounds are usually better than infrared spectra (Diniz et al., 2002). In compounds that present short O-H$\cdots$O HB, two bands in Raman spectra are observed: one around 300 cm^{-1} assigned to symmetric stretching mode of OH$\cdots$O group and around 850 cm^{-1} assigned to the asymmetric one. The O-H stretching mode is observed around 2500 cm^{-1}. In extended structures, the bands assigned as stretching mode of D-H$\cdots$A group (where D is the donor atom of HB and A is the acceptor of HB) appear at 3100 and 2400 cm^{-1} and bands related to deformation modes at 1650-1620 cm^{-1} region and around 1160 cm^{-1}. In solid state investigation of aminopyridinium derivatives (Lorenc et al., 2008a; Lorenc et al., 2008b) the extended HB were identified by single-crystal X-ray diffraction and by Raman and Infrared spectra. Bands related to stretching modes as so as deformation modes of NH$\cdots$O, NH$\cdots$N and OH$\cdots$O groups were observed in vibrational spectra, as can be seen in Table 1. These results confirm the capability of these technics in studies of supramolecular HB in solid state samples.

APBS[a]			AP[a]			ACP[b]			ACPSe[b]			Assignment
IR	R	Calc.	IR	R	Calc.	IR	R	Calc.	IR	R	Calc.	
3351	-	3542	3445	3447	3467	3456	3454	3562	3351	-	3545	v(NH$\cdots$O)
						3297	3298	3444	3288	3317		v_{as}(NH$\cdots$O)
						3233			3243	3191	3175	v_{s}(NH$\cdots$O)
2860								3672	2900-2600	2474	2719 2275	v(OH$\cdots$O)
	1674						1640		1663 1647 1623	1652 1627	1661 1648 1622	δ(NH$\cdots$O)
					1490 1443							δ(NH$\cdots$N)
									1004	1001	1009	γ(O$\cdots$HN)
										450	474	v(O$\cdots$HO)

[a] (Lorenc et al., 2008b), [b] (Lorenc et al., 2008a). APBS = 2-aminopyridinium-4-hydroxybenzenosulfonate, AP = 2-aminopyridine, ACP = 2-amino-5-chloropyridine, ACPSe = 2-amino-5-chloropyridinium hydrogen selenate.

Table 1. Supramolecular hydrogen bond vibrations in aminopyridinium derivatives.

3. Antihypertensive drugs and Raman spectra

Hypertension is one of the most prevalent diseases worldwide affecting millions of people (Brandão et al., 2010). Hypertension occurs when blood pressure levels are above the reference values for the general population. Several groups of drugs are used to combat high blood pressure as the carboxypeptidase enzyme inhibitors and the diuretics.

The chlortalidone (CTD) and hydrochlorothiazide (HYD) diuretics drugs are widely used in hypertension treatment. These drugs are active pharmaceutical ingredients with long-acting oral activity. The CTD has two polymorphic forms described in the literature, form I and form III, respectively (Martins et al., 2009), although the HYD has only one single-crystal structure reported in literature (Dupont and Dideberg, 1972). The enalapril carboxypeptidase enzyme inhibitor drug presents a low solubility which difficult their absorption by human body. Due to this fact the drug is administered as enalapril maleate (ENM), which is a compound more soluble than the enalapril itself. The crystal structure of the ENM drug was described for Précigoux (Precigoux et al., 1986). Another drug very common in hypertension treatment is the losartan potassium (LOS), which is an angiotensin II receptor (type AT1) antagonist (Erk, 2001). In hypertension treatment is common the use of two (or more) drug in association with diuretics. All of these compounds can be used in association and structural and vibrational descriptions of them are show below. The molecular structure and Raman spectra of these drugs are displayed in Figures 2 and 3, respectively. The assignment of the selected bands are listed in Table 2 and marked in Figure 3.

The Raman spectrum of the CTD drug presents vibrational modes of functional groups of CTD molecule, such as stretching modes of NH (3251 cm^{-1}), CO$_{amide}$ (1657 cm^{-1}), SO (1163 cm^{-1}) and CCl (682 cm^{-1}) bonds (Nakamoto, 1986). The v(NH), v(SO), v(CCl) in the Raman spectrum of the HYD drug occur at 3266, 1166 and 610 cm^{-1}, respectively. These bands were used to identify this drug in the associations (Nakamoto, 1986). In ENM drug, bands in the region of 3054 to 2891 cm^{-1} were observed in the Raman spectrum of this compound which can be assigned as v(CH) of aromatic ring, symmetric and asymmetry stretching modes of CH$_2$ and CH$_3$ groups. Other bands as carbonyl stretching of amide, carbonyl stretching of maleate and carbonyl stretching of ester were also observed (Widjaja et al., 2007). For LOS, the most important bands are observed at 807 and 819 cm^{-1} assigned to imidazole ring deformation and at 1489 cm^{-1} attributed to N=N stretching mode (Raghavan et al., 1993).

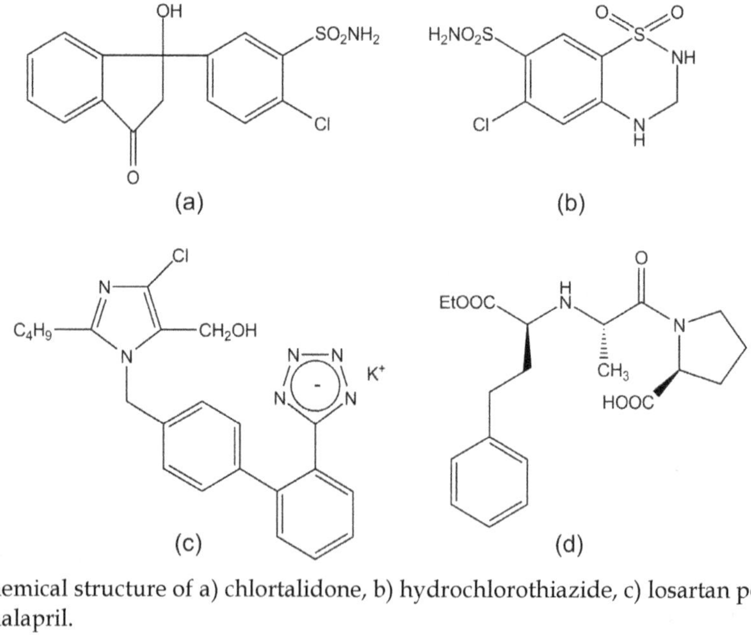

Fig. 2. Chemical structure of a) chlortalidone, b) hydrochlorothiazide, c) losartan potassium and d) enalapril.

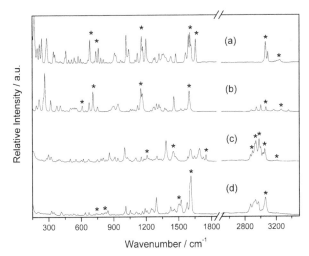

Fig. 3. Raman spectra of a) chlortalidone, b) hydrochlorothiazide, c) enalapril maleate and d) losartan potassium. The * represent the most important bands.

Mode	Compounds			
	CTD	HYD	ENM	LOS
v(CCl)	682	610		
γ(CH)	761	710	751	761
$\Phi_{\text{imidazole ring}}$				810
δ(CH)	1156			
v(SO)	1163	1166		
v(CCO)			1210	
δ_{sym}(CH$_2$)			1451	
v(N=N)				1498
v(CC)	1567	1552		1616
v(CO)$_{\text{maleate}}$			1584	
δ(NH)	1605	1602		
v(CO)$_{\text{amide}}$	1657		1648	
v(CO)$_{\text{enalapril}}$			1730	
v(CO)$_{\text{ester}}$			1750	
v(CH$_3$) and v(CH$_2$)$_{\text{aliphatic}}$			2890-2980	2950-2870
v(CH)	3063	3069	3054	3064
v(NH)	3251	3266	3212	

sym and asym: symmetric and asymmetric modes; δ: in plane bending; γ: out of plane bending.

Table 2. Experimental Raman frequencies for chlortalidone, hydrochlorothiazide, enalapril maleate and losartan potassium drugs.

3.1 Polymorphism and vibrational spectra

In solid pharmaceuticals compounds is very common the occurrence of polymorphism phenomena, which may change physical properties of the compounds; e.g., solubility, melting point, optical and electrical properties, density, hardness and conductivity (Dunitz and Bernstein, 1995). The solubility is one of the most important factors in the drugs absorption, once the modification of this property can change the bioavailability. It may bring serious problems for the patients and for the pharmaceutical industries. For example, in investigation about mebendazole drug (Ferreira et al., 2010), that is used in treatment of worms infestations, three polymorphic forms called A, B and C were identified. The main difference in their physical-chemical properties is the water solubility that modifies the therapeutics effect of these polymorphs. The form A is more soluble and stable than forms B and C. However, the C form is more efficacious in comparison to B form. Other interesting example occur with ritonavir drug, which is a protease inhibitor of human immunodeficiency virus type 1 (HIV-1) and used in the treatment of Acquired Immune Deficiency Syndrome (AIDS) (Chemburkar et al., 2000). A more stable polymorph was reported in the literature (form II), which presents serious solubility problems in compared with the original form (form I). This fact brings problems for patients and pharmaceutical industry, which was forced to remove lots of capsules that had form II, from the market (Bauer et al., 2001). In the literature (Chemburkar et al., 2000) was discussed that the difference between solubility of forms I and II is due to the changes of hydrogen bonds strength in solid state.

Due to this fact, it is very important to identify the presence (or not) of different polymorphic forms in pharmaceuticals formulations. The main technique used in identification of polymorphism is the single-crystal X-Ray diffraction, but due to the difficulty of synthesize this kind of samples, other techniques are used, for example, Raman spectroscopic and X-Ray powder diffraction. Additionally, the Raman spectroscopy may be used in the identification of polymorphism (Raghavan et al., 1993), since different vibration modes can be associated to modifications in molecular packing in crystalline solids. These differences are observed mainly in low-frequency, where may arise lattice vibration that is more sensitive to structural changes in solid state. In this sense, the polymorphism investigation of LOS and CTD will be described in section 3.1.1.

3.1.1 Losartan potassium

Based on aforementioned points an investigation about crystalline phases of antihypertensive LOS by Raman spectroscopy and powder X-Ray is described below. The crystalline forms to LOS drug are described in the literature (Fernandez et al., 2002; Hu et al., 2005b). Figure 4 displays the Raman spectra of two crystalline phases (A and B) for the LOS.

Raman spectrum for sample A is slightly different from sample B spectrum. It can be better see in the region among 950 to 250 cm^{-1} (Figure 4b), where the mainly differences are displayed with asterisk (*). For example, it can be observed that in sample A spectrum the ring breathing mode of the imidazole ring [$\Phi_{imidazole}$] arise at 810 cm^{-1} and the same mode in sample B spectrum appears at 813 cm^{-1}. Apart this, in sample A the band associated to C-H out of plane mode of the biphenyl ring [γ(C-H)] is observed as a shoulder at 761 cm^{-1} and in sample B this band appears at 765 cm^{-1} as a single band.

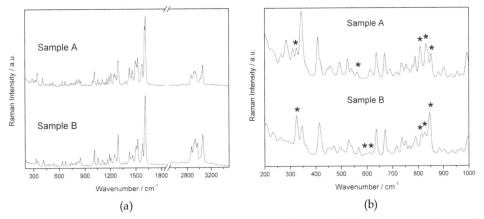

(a)   (b)

Fig. 4. Raman spectra of a) two crystalline phases (samples A and B) losartan potassium and b) a region of 200 to 1000 cm^{-1}.

The X-ray powder diffraction data is in agreement with Raman data, which may be observed in Figure 5, where the samples present different crystalline forms. The peak fitting suggests that the sample A crystallizes in orthorhombic system (Pbca), with unit cell parameters of a = 13.1389(3) Å, b = 25.6885(5) Å, c = 31.1822(7) Å (Hu et al., 2005a). It was observed that the compound is a pseudo-polymorph, due to the presence of the seven water molecules in its unit cell that also contains two potassium cations and two LOS anions. This structure is stabilized by medium and weak hydrogen bond (OH···N and OH···O) (Hu et al., 2005a). For sample B the Bragg`s peak did not fit to any described phases in the literature for this drug. Vibrational and diffraction data of samples A and B suggested that the differences between these two polymorphic forms may be due to differences in intermolecular interaction and crystal symmetry. Thus, the association of Raman spectroscopy and X-ray powder diffraction data can provide a better description of phase analyses in solid pharmaceutical formulations.

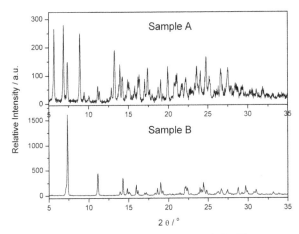

Fig. 5. X-ray powder diffractogram of two samples (A and B) of losartan potassium.

3.1.2 Chlortalidone

The association of computational methodologies and experimental techniques has become in the last few years an essential tool to completely comprehend supramolecular systems (Yan et al., 2007). The wide variety of interactions that rule these systems and their complexity deserve a more detailed analysis in order to describe and to foresee their physical-chemical properties. The computational methods have been rapidly developed in such a way that now it is possible to study systems that some years ago were enviable. For example, Seiffert and co-workers have performed an extended work about the polymorphs structures of lithium-boron imidazolates (Baburin et al., 2011). They have analyzed 30 different structures concerning the potential capability of hydrogen storage of these species.

The effect of the polymorphism in pharmacological science is quite important in the bioavailability, since it can affect physical properties such as solubility and stability of the compound (Yamada et al., 2011). The effects of the polymorphic form on the performance, stability and efficacy of the pharmacological forms have been widely investigated (Brittain, 2009, 2010). The CTD (Figure 2a) presents three different polymorphs forms, but only two of them already have their crystallographic structure known (forms I and III). In this way we have performed density functional theory (DFT) calculations in order to analyze these two distinct species of CTD.

The theoretical investigation of the polymorphs structures of CTD was carried out with periodic density functional theory (DFT) calculations employing the LDA approximation according to Perdew-Zunger (PZ) functional. The wave functions of valence electrons are shown by a plane wave basis set with maximum kinetic energy of 30 hartree (60 Ry). The optimization process was taken using a grid with 2 x 2 x 2 k points. The starting geometries for those calculations were the crystallographic data (Martins et al., 2009). All the calculations were performed in the Quantum Espresso (Baroni et al.) (PWscf) program package. The geometries were fully optimized including the atomic coordinates and the lattice parameters. The calculations of vibrational frequencies at the Γ point, were performed on the optimized geometry. These vibrational frequencies are quite useful since they can be used to help in the experimental attributions of the bands in infrared spectrum (De Abreu et al., 2009).

In Figure 6 are shown the optimized structures for the CTD polymorphs I (Fig. 6a) and III (Fig. 6b). It is interesting to note in these figures that both structures are stabilized through hydrogen bonds forming a network in three dimensions. In form I it can be observed the formation of a dimmeric structure in an 8 membered ring through the $NH \cdots OC/OC \cdots HN$ atoms. In this same polymorph there is another hydrogen bond formed between the $OH \cdots OC$ groups. In form III there are two different hydrogen bonds stabilizing the supramolecular structure, one involving the groups $OH \cdots OC$, like in form I, and another with the $NH_2 \cdots OC$ groups, that it is not present in form I. Table 3 contains the experimental and theoretical crystal data for both forms. One can note that there was a contraction in the estimated crystal axis of both polymorphs compared to experimental, the difference is about 2 – 7%. The angles are also well described presenting an error bar of -2 – 1% that represents an expansion and contraction, respectively. The crystallographic parameter that suffered the biggest deviance was the unit cell volume that showed an error about 10 and 12% to forms I and III, respectively. However these changes in the crystallographic parameters did not modify the symmetry of the solid (monoclinic system).

	Form I		Form III	
	Experimental	Calculated	Experimental	Calculated
a / Å	6.2270(2)	6.0227	7.9957(2)	7.8520
b / Å	8.3870(3)	8.1662	8.1467(2)	7.7988
c / Å	14.3640(4)	13.8537	11.4761(3)	10.7156
α / °	92.141(2)	91.6952	80.448(2)	80.861
β / °	101.050(2)	101.583	79.277(2)	79.527
γ / °	107.024(2)	108.088	86.106(2)	87.652
V / Å³	700.50(4)	631.40	723.77(3)	637.02

[a]Experimental data taken from (Martins et al., 2009). [b]Theoretical values calculated in this work.

Table 3. Experimental[a] and theoretical[b] crystal data for chlortalidone forms I and III.

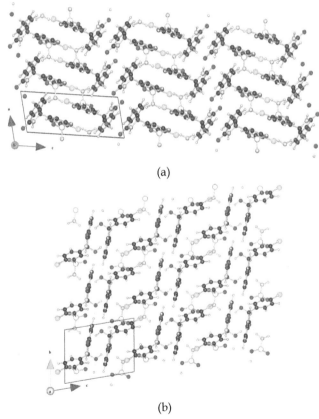

(a)

(b)

Fig. 6. Optimized structures of the chlortalidone polymorphs I (a) and III (b).

Concerning the energetics of these species, form I is about 84 kcal mol^{-1} more stable than form III, if we taking into account only the electronic energy. This energetic difference is probably due to the presence of more intermolecular interactions in form I than in III. Table 4 presents the experimental and theoretical infrared attribution for the polymorphs I and III of CTD. The main vibrational modes related to the intramolecular hydrogen bonds were chosen in a tentative to distinguish the two species. In form I the N-H group is involved in the dimmer formation through a hydrogen bond interaction, it is evidenced by the shift to higher wavenumbers, from 3000 cm^{-1} to 3181 cm^{-1} of the stretching mode of N-H group. The same behavior is observed in the v (O-H) mode that is shifted from 2861 cm^{-1} to 2968 cm^{-1}, as can be seen in table 4. The formation of intermolecular hydrogen bonds in form III can be evidenced if one compares the shift in the v (O-H) and v (NH $\cdots$ O) calculated modes. It is important to mention that in high frequency is observed the biggest deviations from the experimental values.

Mode	Experimental Form I	Theoretical Form I	Form III
v(S-NH$_2$)	918	931	946
v(S-O)	1165	1154	1276
v(N-H)	3259	3000	3181
v(O-H)	2858	2861	2968
v(C=O)	1681	1670	1647
v(C-OH)	1043	1042	1063
v(N-H$_2$)	3259	3200	3110 / 3117
v(NH...O)	3550		
v(NH...O)as	3280	2942, 2998	3008 / 3031
v(NH...O)s	3240		
v(OH...O)	2720	2861	2968 / 2988
δ (NH...O)	1600	1628	1635
δ (O...HN)	1000	-	-

Table 4. Experimental (form I) and theoretical infrared attributions of the chlortalidone forms I and III. Wavenumbers are in cm^{-1}.

3.2 Association of hypertensive drugs
3.2.1 Lorsatan potassium in association with chlortalidone and hydrochlorothiazide
To the bands identification, these pharmaceuticals were mixed in a stoichiometric proportion (1:1), (1:2), (4:1), (2:1). It is important emphasize that the calculation of the different proportions was made using the minimum dosage of the associations that is 50 mg for LOS and 12.5 mg for diuretics. Before analyses of these associations, it is necessary know the main Raman bands of the diuretics, since they are used in minimum dosage in pharmaceuticals associations when compared with LOS.
Figures 7 and 8 display the characteristic bands of the LOS/CTD and LOS/HYD associations in the different proportions. It can be observed in the (1:1) rate characteristic bands of the CTD, Figure 7a, at 3094 and 1657 cm^{-1}. The same bands were observed in the rates (1:2) and (1:4). Moreover for (1:2) and (1:4) can be also observed the bands at 3250, 1474, 909, 759, 682 and 460 cm^{-1}. The difference is due to the fact that relative intensity in 1:4 is bigger than in 1:2. However, when for association where LOS increase, 2:1 rate, only one band assigned to CTD drug is verified at 1657 cm^{-1} (Figure 7d). On the other hand, in the 4:1 rate (Figure 7e) the profile of the spectrum is characteristic of LOS drug.

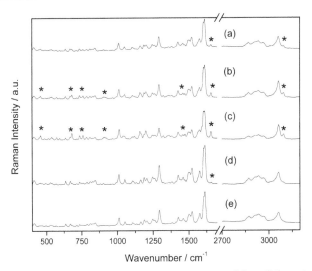

Fig. 7. Raman spectra of losartan potassium in association to chlortalidone in proportions: a) 1:1, b) 1:2, c) 1:4, d) 2:1 and e) 4:1.

Similar results are observed for association with LOS and HYD. Figure 8 shows the Raman spectra of this association. It can be observed that in the rate 1:1 (Figure 8a), just one band was assigned to the presence of HYD at 3005 cm^{-1}. For 1:2 and 1:4 rates (Figures 8b and 8c) were verified bands at 3361, 3268, 3170, 3005, 1320, 1152, 902 and 601 cm^{-1}, characteristics of the HYD. The main different between them is that in 1:4 rate, the relative intensity is bigger than 1:2 rate. In the 2:1 rate the characteristic band of the HYD was observed at 3005 cm^{-1} (Figure 8d) and for the 4:1 proportion the entire spectrum is characteristic of the LOS (Figure 8e).

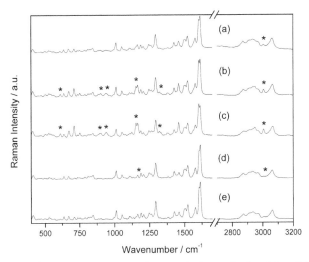

Fig. 8. Raman spectra of losartan potassium in association to hydrochlorothiazide in proportions: a) 1:1, b) 1:2, c) 1:4, d) 2:1 and 4:1.

The optimization of this fast and simple method of analyze is important, mainly in antihypertensive drugs associations. These results show the potential of Raman spectroscopy in qualitative analysys in the pharmaceutical associations, however when the LOS is in large proportions, like in 4:1 rate, occur the overlapping between the bands of pharmaceutical formulations and it is not possible characterized the diuretics bands.

3.2.2 Enalapril maleate in association with chlortalidone and hydrochlorothiazide

The Raman spectra of the ENM/CTD and ENM/HYD associations are shown in Figures 9 to 14. The proportions of the drugs at the associations were 1:2.5, 1.6:1 and 2:1. The proportions respected the minimum dosage of each drug what is 12.5 mg for CTD and HYD and 5 mg for ENM (Brandão et al., 2010).

Figure 9c shows the Raman spectrum of the ENM/CTD association in proportion 1:2.5 (1 mg of ENM for 2.5 mg of CTD). In this Raman spectrum all characteristic bands of the CTD drug are present and have relative intensities larger those characteristic bands of the ENM drug. This is due to the large quantity of the CTD drug in the mixture. Some bands of the ENM are not present in the Raman spectra of the association obtained.

In Raman spectrum of the ENM and CTD association in proportion 1.6:1 (1.6 mg of ENM for 1 mg of CTD) shown in Figure 10c are observed characteristic bands of the ENM and CTD drugs. The relative intensities of the characteristic bands of CTD drug were reduced due to lower proportion of this drug in the mixture.

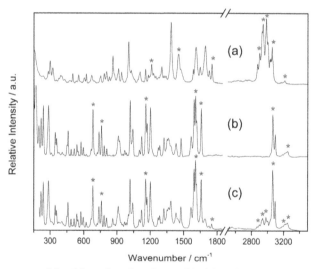

Fig. 9. Raman spectra of the (a) enalapril maleate, (b) chlortalidone and (c) enalapril maleate and chlortalidone association in proportion 1:2.5. The symbol * refers to bands relative to enalapril maleate vibrations and * the bands relative to chlortalidone vibrations.

Figure 11c shows the Raman spectrum of the ENM and CTD association in proportion 2:1 (2 mg of ENM for 1 mg of CTD). This spectrum is very similar to that showed in Figure 10c. The characteristic bands of the ENM and CTD drugs are observed being that the relative intensities of the CTD bands are smaller than ENM. This similarity between these spectra refers to almost equal proportion of drugs in the mixtures.

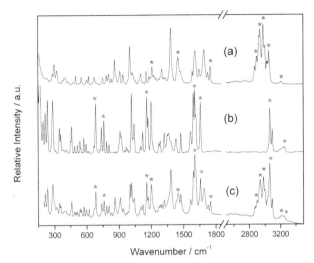

Fig. 10. Raman spectra of the (a) enalapril maleate, (b) chlortalidone and (c) enalapril maleate and chlortalidone association in proportion 1.6:1. The symbol * refers to bands relative to enalapril maleate vibrations and * the bands relative to chlortalidone vibrations.

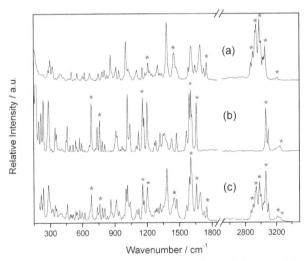

Fig. 11. Raman spectra of the (a) enalapril maleate, (b) chlortalidone and (c) enalapril maleate and chlortalidone association in proportion 2:1. The symbol * refers to bands relative to enalapril maleate vibrations and * the bands relative to chlortalidone vibrations.

Figure 12c shows the Raman spectrum of the ENM/HYD association in proportion 1:2.5 (1 mg of ENM for 2.5 mg of HYD). In this Raman spectrum all characteristic bands of the HYD drug are present and have relative intensities larger than the characteristic bands of the ENM drug. Similar to the ENM/CDT, this is due to the large quantity of the HYD drug in the mixture. Some bands of the ENM are not present in the Raman spectra of associations.

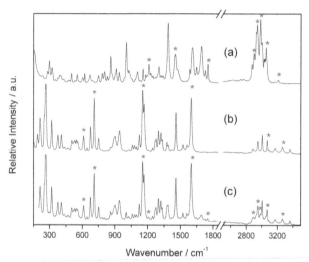

Fig. 12. Raman spectra of the (a) enalapril maleate, (b) hydrochlorothiazide and (c) enalapril maleate and hydrochlorothiazide association in proportion 1:2.5. The symbol * refers to bands relative to enalapril maleate vibrations and * the bands relative to hydrochlorothiazide vibrations.

In Raman spectrum of the ENM/HYD association in proportion 1.6:1 (1.6 mg of ENM for 1 mg of HYD) shown in Figure 13c are observed characteristic bands of the ENM and HYD drugs. The relative intensities of the characteristic bands of HYD drug such as $v(NH)$, $v(SO)$ and $\gamma(CH)$ were reduced due to the lower proportion of this drug in the mixture.

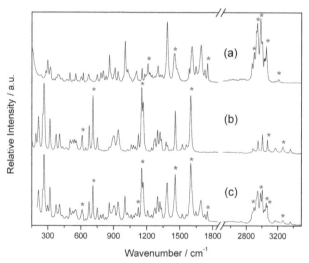

Fig. 13. Raman spectra of the (a) enalapril maleate, (b) hydrochlorothiazide and (c) enalapril maleate and hydrochlorothiazide association in proportion 1.6:1. The symbol * refers to bands relative to enalapril maleate vibrations and * the bands relative to hydrochlorothiazide vibrations.

Figure 14c shows the Raman spectrum of the ENM and HYD association in proportion 2:1 (2 mg of ENM for 1 mg of HYD). The characteristic bands of the ENM and HYD drugs are observed being that the relative intensities of HYD bands are smaller than ENM band intensities.

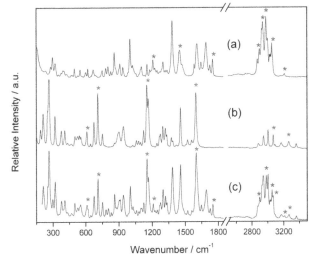

Fig. 14. Raman spectra of the (a) enalapril maleate, (b) hydrochlorothiazide and (c) enalapril maleate and hydrochlorothiazide association in proportion 2:1. The symbol * refers to bands relative to enalapril maleate vibrations and * the bands relative to hydrochlorothiazide vibrations.

4. Conclusion

These results show the potential of Raman spectroscopy in the identification of polymorphism in pharmaceuticals as well in the qualitative analyze in drugs associations. Additionally, the combination of these results with X-ray diffraction and theoretical calculations data complete the description of the solid state, in special in supramolecular chemistry investigations.

5. Acknowledgment

The authors are thankful to the Brazilian Agencies CAPES, FAPEMIG, CNPq for the financial support, LabCri (Instituto de Física – Universidade Federal de Minas Gerais) and LDRX (Instituto de Física – Universidade Federal Fluminense) for X-ray facilities, and Filipe B. de Almeida and Chris H. J. Franco for help in infrared and Raman spectra.

6. References

Aaltonen, J., Gordon, K.C., Strachan, C.J., Rades, T. (2008). Perspectives in the use of spectroscopy to characterise pharmaceutical solids. *International Journal of Pharmaceutics*, Vol. 364, No 2, pp. (159-169), 0378-5173

Baburin, I.A., Assfour, B., Seifert, G., Leoni, S. (2011). Polymorphs of lithium-boron imidazolates: energy landscape and hydrogen storage properties. *Dalton Transactions*, Vol. 40, No 15, pp. (3796-3798), 1477-9226

Baroni, S., dalCorso, A., deGironcoli, S., Giannozzi, P., Cavazzoni, C. *http://www.democritos.it*

Bauer, J., Spanton, S., Henry, R., Quick, J., Dziki, W., Porter, W., Morris, J. (2001). Ritonavir: An Extraordinary Example of Conformational Polymorphism. *Pharmaceutical Research*, Vol. 18, No 6, pp. (859-866), 0724-8741

Brandão, A.A., Magalhães, M.E.C., Ávila, A., Tavares, A., Machado, C.A., Campana, E.M.G., Lessa, I., Krieger, J.E., Scala, L.C., Neves, M.F., Silva, R.C.G., Sampaio, R. (2010). VI Diretrizes Brasileiras de Hipertensão. *Arquivos Brasileiros de Hipertensão*, Vol. 95, No (11 - 17), 0066-782X

Brittain, H.G. (2009). Polymorphism and Solvatomorphism 2007. *Journal of Pharmaceutical Sciences*, Vol. 98, No 5, pp. (1617-1642), 0022-3549

Brittain, H.G. (2010). Polymorphism and Solvatomorphism 2008. *Journal of Pharmaceutical Sciences*, Vol. 99, No 9, pp. (3648-3664), 0022-3549

Carlucci, L., Ciani, G., Proserpio, D.M. (2003). Polycatenation, polythreading and polyknotting in coordination network chemistry. *Coordination Chemistry Reviews*, Vol. 246, No 1-2, pp. (247-289), 0010-8545

Chagas, L.H., Janczak, J., Gomes, F.S., Fernandes, N.G., de Oliveira, L.F.C., Diniz, R. (2008). Intermolecular interactions investigation of nickel(II) and zinc(II) salts of ortho-sulfobenzoic acid by X-ray diffraction and vibrational spectra. *Journal of Molecular Structure*, Vol. 892, No 1-3, pp. (305-310), 0022-2860

Chemburkar, S.R., Bauer, J., Deming, K., Spiwek, H., Patel, K., Morris, J., Henry, R., Spanton, S., Dziki, W., Porter, W., Quick, J., Bauer, P., Donaubauer, J., Narayanan, B.A., Soldani, M., Riley, D., McFarland, K. (2000). Dealing with the Impact of Ritonavir Polymorphs on the Late Stages of Bulk Drug Process Development. *Organic Process Research & Development*, Vol. 4, No 5, pp. (413-417), 1083-6160

De Abreu, H.A., Junior, A.L.S., Leitao, A.A., De Sa, L.R.V., Ribeiro, M.C.C., Diniz, R., de Oliveira, L.F.C. (2009). Solid-State Experimental and Theoretical Investigation of the Ammonium Salt of Croconate Violet, a Pseudo-Oxocarbon Ion. *Journal of Physical Chemistry A*, Vol. 113, No 23, pp. (6446-6452), 1089-5639

Diniz, R., de Abreu, H.A., de Almeida, W.B., Sansiviero, M.T.C., Fernandes, N.G. (2002). X-ray crystal structure of triaquacopper(II) dihydrogen 1,2,4,5-benzenetetracarboxylate trihydrate and Raman spectra of Cu2+Co2+, and Fe2+ salts of 1,2,4,5-benzenetetracarboxylic (pyromellitic) acid. *European Journal of Inorganic Chemistry*, Vol., No 5, pp. (1115-1123), 1434-1948

Dunitz, J.D., Bernstein, J. (1995). Disappearing Polymorphs. *Accounts of Chemical Research*, Vol. 28, No 4, pp. (193-200), 0001-4842

Dupont, L., Dideberg, O. (1972). Structure cristalline de l'hydrochlorothiazide, C7H8ClN3O4S2. *Acta Crystallographica Section B*, Vol. 28, No 8, pp. (2340-2347), 0567-7408

Emsley, J. (1980). VERY STRONG HYDROGEN-BONDING. *Chemical Society Reviews*, Vol. 9, No 1, pp. (91-124), 0306-0012

Erk, N. (2001). Analysis of binary mixtures of losartan potassium and hydrochlorothiazide by using high performance liquid chromatography, ratio derivative spectrophotometric and compensation technique. *Journal of Pharmaceutical and Biomedical Analysis*, Vol. 24, No 4, pp. (603-611), 0731-7085

Fernandez, D., Vega, D., Ellena, J.A., Echeverria, G. (2002). Losartan potassium, a non-peptide agent for the treatment of arterial hypertension. *Acta Crystallographica Section C-Crystal Structure Communications*, Vol. 58, No (m418-m420), 0108-2701

Ferreira, F.F., Antonio, S.G., Rosa, P.C.P., Paiva-Santos, C.d.O. (2010). Crystal structure determination of mebendazole form A using high-resolution synchrotron x-ray powder diffraction data. *Journal of Pharmaceutical Sciences*, Vol. 99, No 4, pp. (1734-1744), 1520-6017

Gonzalezsanchez, F. (1958). INFRA-RED SPECTRA OF THE BENZENE CARBOXYLIC ACIDS. *Spectrochimica Acta*, Vol. 12, No 1, pp. (17-33),

Hoeben, F.J.M., Jonkheijm, P., Meijer, E.W., Schenning, A. (2005). About supramolecular assemblies of pi-conjugated systems. *Chemical Reviews*, Vol. 105, No 4, pp. (1491-1546), 0009-2665

Hu, X.-R., Wang, Y.-W., Gu, J.-M. (2005a). Losartan potassium 3.5-hydrate, a new crystalline form. *Acta Crystallographica Section E*, Vol. 61, No 9, pp. (m1686-m1688), 1600-5368

Hu, X.R., Wang, Y.W., Gu, J.M. (2005b). Losartan potassium 3.5-hydrate, a new crystalline form. *Acta Crystallographica Section E-Structure Reports Online*, Vol. 61, No (M1686-M1688), 1600-5368

Kollman, P.A., Allen, L.C. (1972). THEORY OF HYDROGEN-BOND. *Chemical Reviews*, Vol. 72, No 3, pp. (283-&), 0009-2665

Lorenc, J., Bryndal, I., Marchewka, M., Kucharska, E., Lis, T., Hanuza, J. (2008a). Crystal and molecular structure of 2-amino-5-chloropyridinium hydrogen selenate - its IR and Raman spectra, DFT calculations and physicochemical properties. *Journal of Raman Spectroscopy*, Vol. 39, No 7, pp. (863-872), 0377-0486

Lorenc, J., Bryndal, I., Marchewka, M., Sasiadek, W., Lis, T., Hanuza, J. (2008b). Crystal and molecular structure of 2-aminopyridinium-4-hydroxybenzenosulfonate - IR and Raman spectra, DFT calculations and physicochemical properties. *Journal of Raman Spectroscopy*, Vol. 39, No 5, pp. (569-581), 0377-0486

Martins, F.T., Bocelli, M.D., Bonfilio, R., de Araujo, M.B., de Lima, P.V., Neves, P.P., Veloso, M.P., Ellena, J., Doriguetto, A.C. (2009). Conformational Polymorphism in Racemic Crystals of the Diuretic Drug Chlortalidone. *Crystal Growth & Design*, Vol. 9, No 7, pp. (3235-3244), 1528-7483

Nakamoto, K. (1986). *Infrared and Raman Spectra of Inorganic and Coordination Compounds* (4th), John Wiley & Sons, 0471010669, New York

Precigoux, G., Geoffre, S., Leroy, F. (1986). N-(1-Ethoxycarbonyl-3-phenylpropyl)-l-alanyl-l-prolinium-hydrogen maleate (1/1), enalapril (MK-421). *Acta Crystallographica Section C*, Vol. 42, No 8, pp. (1022-1024), 0108-2701

Raghavan, K., Dwivedi, A., Campbell Jr, G.C., Johnston, E., Levorse, D., McCauley, J., Hussain, M. (1993). A Spectroscopic Investigation of Losartan Polymorphs. *Pharmaceutical Research*, Vol. 10, No 6, pp. (900-904), 0724-8741

Sobczyk, L., Grabowski, S.J., Krygowski, T.M. (2005). Interrelation between H-bond and Pi-electron delocalization. *Chemical Reviews*, Vol. 105, No 10, pp. (3513-3560), 0009-2665

Takahashi, O., Kohno, Y., Nishio, M. (2010). Relevance of Weak Hydrogen Bonds in the Conformation of Organic Compounds and Bioconjugates: Evidence from Recent

Experimental Data and High-Level ab Initio MO Calculations. *Chemical Reviews*, Vol. 110, No 10, pp. (6049-6076), 0009-2665

Wang, K., Duan, D., Wang, R., Liu, D., Tang, L., Cui, T., Liu, B., Cui, Q., Liu, J., Zou, B., Zou, G. (2009). Pressure-Induced Phase Transition in Hydrogen-Bonded Supramolecular Adduct Formed by Cyanuric Acid and Melamine. *Journal of Physical Chemistry B*, Vol. 113, No 44, pp. (14719-14724), 1520-6106

Widjaja, E., Lim, G.H., Chow, P.S., Tan, S. (2007). Multivariate data analysis as a tool to investigate the reaction kinetics of intramolecular cyclization of enalapril maleate studied by isothermal and non-isothermal FT-IR microscopy. *European Journal of Pharmaceutical Sciences*, Vol. 32, No 4-5, pp. (349-356), 0928-0987

Yamada, H., Masuda, K., Ishige, T., Fujii, K., Uekusa, H., Miura, K., Yonemochi, E., Terada, K. (2011). Potential of synchrotron X-ray powder diffractometry for detection and quantification of small amounts of crystalline drug substances in pharmaceutical tablets. *Journal of Pharmaceutical and Biomedical Analysis*, Vol. 56, No 2, pp. (448-453), 0731-7085

Yan, S., Lee, S.J., Kang, S., Lee, J.Y. (2007). Computational approaches in molecular recognition, self-assembly, electron transport, and surface chemistry. *Supramolecular Chemistry*, Vol. 19, No 4-5, pp. (229-241), 1061-0278

Ye, B.H., Tong, M.L., Chen, X.M. (2005). Metal-organic molecular architectures with 2.2 '-bipyridyl-like and carboxylate ligands. *Coordination Chemistry Reviews*, Vol. 249, No 5-6, pp. (545-565), 0010-8545

5

Vibrational and Optical Studies of Organic Conductor Nanoparticles

Dominique de Caro, Kane Jacob, Matthieu Souque and Lydie Valade

CNRS, LCC (Laboratoire de Chimie de Coordination), 205, route de Narbonne
and Université de Toulouse, UPS, INPT, LCC
Toulouse,
France

1. Introduction

To create free electrons in organic solids and thus generate an organic material exhibiting electrical conductivity, a simple way is to build an organic complex, in which there is a charge transfer from the atoms or molecules of an electron donor (D) to those of an electron acceptor (A). In 1973, the charge transfer salt TTF·TCNQ (donor: tetrathiafulvalene, TTF; acceptor: tetracyanoquinodimethane, TCNQ) was synthesized (figure 1) (Ferraris et al., 1973). In TTF·TCNQ single crystals, TTF and TCNQ form segregated columnar stacks along the *b*-axis of the crystal structure. The interplanar spacings in the TTF and TCNQ molecular stacks at room temperature are 3.47 Å and 3.17 Å, respectively (Kistenmacher et al., 1974). The TTF and TCNQ molecular planes tilt at an angle of 24.5 ° and 34.0 °, respectively with respect to the *b*-axis forming a herringbone arrangement. As single crystals, this compound behaves like a metal (the dc conductivity increases with decreasing temperature) down to 54 K, temperature at which it undergoes a metal-to-semiconductor transition. The maximum of conductivity in TTF·TCNQ is along the *b*-axis (about 600 Ω^{-1} cm^{-1}). Conductivity values range from 10^{-2} to 1 Ω^{-1} cm^{-1} in a direction perpendicular to the direction of maximum conductivity. The amount of charge transfer from the TTF donor molecule to the TCNQ acceptor molecule has been investigated by various techniques. Using X-ray photoelectron spectroscopy, a value of 0.56 ± 0.05 has been extracted from the shape of the S2p signal (Ikemoto et al., 1977). Values in the range 0.50–0.60 have been obtained by a numerical integration of X-ray diffraction amplitudes (Coppens, 1975). However, the most convenient technique is based on infrared spectroscopy. Using the linear correlation of the nitrile stretching mode for TCNQ as a function of the degree of charge transfer, a value of 0.59 ± 0.01 has been obtained (Chappell et al., 1981).

Although synthesized for many years, TTF·TCNQ currently attracts much interest because of its interesting physical properties. Moreover, it is the one-dimensional conductor which is the most intensively processed in forms others than single crystals. For instance, TTF·TCNQ was prepared as thin films on (100)-oriented alkali halide substrates (Fraxedas et al., 2002), as self-organized monolayers on Au(111) (Yan et al., 2009), or as nanowires on stainless steel conversion coatings (Savy et al., 2007). We report in this chapter the preparation and spectral studies of TTF·TCNQ prepared as nanoparticles.

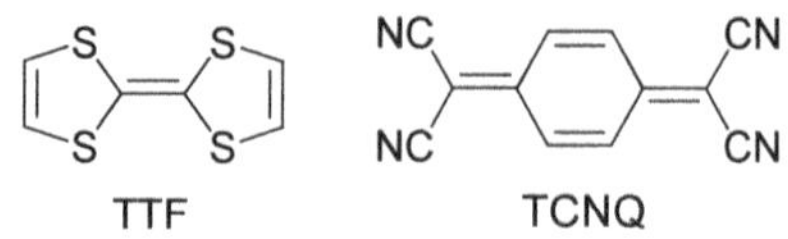

Fig. 1. Molecular structures for TTF and TCNQ

2. Nanoparticle synthesis

The TTF·TCNQ charge transfer salt is prepared by slow diffusion of an organic solution of TTF into an organic solution of TCNQ. Due to its quasi-one-dimensional character, TTF·TCNQ is commonly grown as long needles or wires. To control the growth of this material as nanospheres or nanoplatelets, a stabilizing agent is added. The stabilizing agent forms a very thin protective layer around each particle thereby preventing their aggregation. We have used ionic liquids (IL), ionic liquid/oleic acid (OA) mixtures or various other as protecting agents or stabilizing media.

Pure ionic liquids such as imidazolium salts are known to stabilize metal nanoparticles (Ru or Pd for example) exhibiting sizes lower than 10 nm (Gutel et al., 2007). The imidazolium salt alone is not suitable for the growth of TTF·TCNQ nanoparticles because precursors are not soluble in this medium. The use of a co-solvent (*e.g.* acetonitrile or acetone, noted S) is essential. For this reason, we have performed the synthesis in a binary (IL/S) or a tertiary (IL/OA/S) mixture. The ionic liquid which has been used as a protecting species is either the 1-butyl-3-methylimidazolium tetrafluoroborate (BMIMBF$_4$, figure 2) or the 1-decyl-3-methylimidazolium tetrafluoroborate (DMIMBF$_4$). The solvent (S) is a 1:1 (vol./vol.) acetonitrile/acetone mixture.

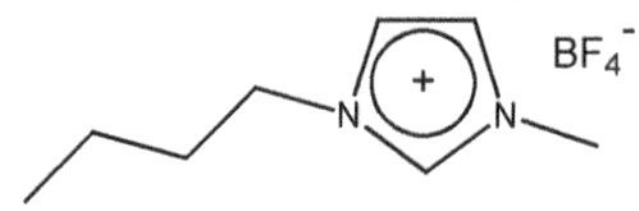

Fig. 2. Molecular structure for BMIMBF$_4$ (for DMIMBF$_4$, the butyl chain C$_4$H$_9$ is replaced by a decyl chain C$_{10}$H$_{21}$)

V$_r$	BMIMBF$_4$	DMIMBF$_4$
0.04	Nanoparticles slightly agglomerated (diameter: 2–6 nm; mean: 3.8 nm)	Mixture of spherical nanoparticles (3–18 nm), elongated nanoparticles, and nano-platelets
0.2–0.4	Well-dispersed nanoparticles (diameter: 12–62 nm; mean: 35 nm)	Well-dispersed nanoparticles (diameter: 30–100 nm; mean: 50 nm)
≥ 1	Long needles (> 5 µm long; 0.5–2 µm wide)	Long needles (> 5 µm long; 0.5–2 µm wide)

Table 1. Transmission electrons microscopy results for TTF·TCNQ prepared in the presence of an ionic liquid

The TTF precursor is solubilized in a mixture of BMIMBF$_4$ (or DMIMBF$_4$) and S whereas the TCNQ precursor is solubilized in S. The TTF solution is slowly added to the TCNQ solution at room temperature. A fine black precipitate appears throughout the addition. The air-stable black solid, filtered off, washed and finally dried under vacuum, consists in TTF·TCNQ nanoparticles (yield ~ 50–80 %). Mean size, morphology and state of dispersion of the nanoparticles depend on the volume ratio V_r = BMIMBF$_4$/S (or DMIMBF$_4$/S), see table 1 and figures 3 and 4. The use of a tertiary mixture (BMIMBF$_4$/OA/S) in a 1:1 ratio (vol./vol.) BMIMBF$_4$/OA leads to clusters of nanoparticles (cluster size: 150–400 nm, size of individual particles in a cluster: 8–25 nm). For a 1:3 ratio (vol./vol.) BMIMBF$_4$/OA, well dispersed nanoparticles exhibiting a mean diameter of 40 nm are obtained.

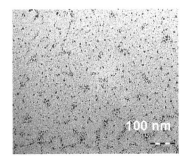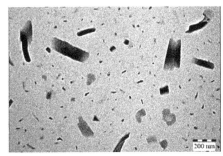

Fig. 3. Electron micrographs for TTF·TCNQ in the presence of BMIMBF$_4$ (left, V_r = 0.04) and DMIMBF$_4$ (right, V_r = 0.04)

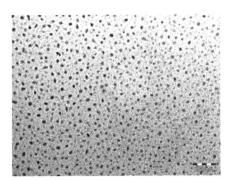

Fig. 4. Electron micrograph for TTF·TCNQ in the presence of BMIMBF$_4$ (V_r = 0.4); bar = 200 nm

3. Optical properties of TTF·TCNQ nanoparticles

The electronic properties of one-dimensional systems are governed by three types of interactions among the unpaired electrons occupying the highest molecular orbital in the solid. These interactions are (i) the overlap of the wave functions of these electrons between adjacent sites in the crystal, (ii) the interactions of the electrons with their surroundings (e.g. phonons), and (iii) the Coulomb interaction between electrons. Theoretical models taking into account interactions (i) are usually based on a tight-binding method for the band structure. The energy bands are of the form:

$$E\left(k\right) = \pm\,2\,t\cos\left(k\,a\right) + \text{const.} \tag{1}$$

where k is the wave vector, a the lattice constant, and t is the transfer matrix element between sites, given by:

$$t = \left\langle \phi_i \left| H \right| \phi_{i+1} \right\rangle \tag{2}$$

where H is the total Hamiltonian of the systems, and ϕ_i is the appropriate molecular orbital wave function.

The width of the band, w = 4t, depends upon the overlap of electronic wave functions, and a one-dimensional character for the bands is obtained by allowing this overlap to occur only in one direction. For partially filled bands, the system will be a metal, with a complex dielectric function:

$$\varepsilon \left(\omega \right) = \varepsilon_\infty - \frac{\omega_p^2}{\omega^2 + i \frac{\omega}{\tau}} \tag{3}$$

where ε_∞ is the dielectric constant at high frequency arising from core polarisability, τ is the electron relaxation time and ω_p is the plasma frequency. For typical one-dimensional charge-transfer based organic conductors, it has been found that $\varepsilon_\infty \sim 3$, $\tau \sim 10^{-15}$ s, and $\omega_p \sim 10000$ cm^{-1}. Several peaks (undoubtedly characterizing a charge-transfer-based organic conductor) can be seen on electronic spectra recorded either in solution or in the solid state. In addition to the plasma frequency, four absorption bands are usually observed: the first one is at $(2–4)\times10^3$ cm^{-1}, the second at about $(10–12)\times10^3$ cm^{-1}, the third one at about $(16–18)\times10^3$ cm^{-1}, and the fourth one at about $(25–31)\times10^3$ cm^{-1}. The first one is assigned to the charge transfer of the type $A^-A^0 \rightarrow A^0A^-$, where A^- and A^0 denote the anion and the neutral molecule of the electron acceptor, respectively (CT1 band). The second one is attributed to the charge transition of the type $A^-A^- \rightarrow A^0A^{2-}$ (CT2 band). The third one is the local excitation associated with the lowest intramolecular transition of A^- (LE1 band), whereas the fourth one is due to the local excitation associated with the lowest intramolecular transition of A^0 (LE2 band).

The room temperature reflectance spectrum of a nanoparticle film of TTF·TCNQ has been recorded in the 9000–25000 cm^{-1} range (figure 5). The general shape of this spectrum is in good agreement with that of TTF·TCNQ single crystals, recorded for an electric field polarized parallel to the b crystallographic axis (Grant et al., 1973). The plasma reflection is clearly seen at 10700 cm^{-1} (about 1.32 eV), in excellent agreement with that obtained for TTF·TCNQ single crystals (1.38 eV) (Grant et al., 1973). The conduction-band width (w = 4t) can be estimated from the measured plasma frequency, using the expression:

$$w = \frac{h^2 \omega_p^2}{16 \pi^2 N e^2 b} \tag{4}$$

where N denotes the electron density and b the lattice parameter (3.819 Å). Taking the value of 4.7×10^{21} cm^{-3} for N (Bright et al., 1974), we find w = 0.59 eV, in good agreement with the w value calculated for TTF·TCNQ single crystals (0.62 eV) (Graja, 1997). It is to be noticed that the X-ray diffraction pattern of the nanoparticle film is dominated by the (002), (004), and (008) lines. This indicates a preferential orientation of the film, $i.e.$ the ab plane being parallel to the substrate surface. In our case, the light is unpolarized and interacts with the substrate surface according to all possible angles (integrating sphere). However, our reflectance spectrum is as good as that recorded for a light polarized parallel to the *b* axis.

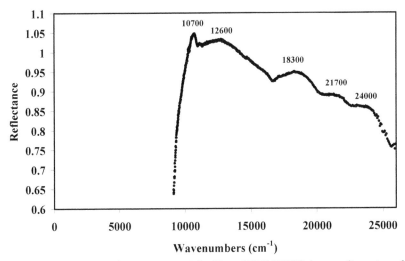

Fig. 5. Reflectance spectrum for a nanoparticle film of TTF·TCNQ (mean diameter of individual particles: 35 nm)

Furthermore, four broads signals are observed at 12600 (CT2), 18300 (LE1), 21700, and 24000 (LE2) cm⁻¹ (figure 5). These four signals are also observed in the reflectance spectrum of TTF·TCNQ single crystals (12000, 17000, 22000, and 25800 cm⁻¹) (Grant et al., 1973). The absorption spectrum of TTF·TCNQ nanoparticles dispersed in acetonitrile (figure 6) also clearly evidences CT2, LE1, and LE2 bands at 11900, 16400–19000, and 26000 cm⁻¹, respectively

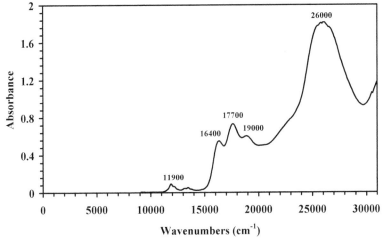

Fig. 6. Absorption spectrum for TTF·TCNQ nanoparticles dispersed in acetonitrile solution (mean diameter: 35 nm)

4. Vibrational properties of TTF·TCNQ nanoparticles

Vibrational spectroscopy (infrared and Raman) is a powerful tool to investigate one-dimensional organic charge-transfer complexes. At large distances between the donor (D) and the acceptor (A) molecules, the vibrational spectrum of D + A is just the sum of those of the free molecules. As the distance between them becomes shorter (formation of the D–A complex), the electrostatic field of one molecule begins to influence the second one, and vice-versa. This causes changes in the frequency and intensity from the spectra of isolated molecules. Moreover, due to the fact that the D (or A) molecule may have a lower symmetry in D–A, additional frequencies appear in the complex because forbidden modes may become active due to mixing with other internal modes of D (or A). The vibrational spectrum consists of a primary electronic charge transfer band (CT1) and a series of oscillations driven by totally symmetric internal molecular vibrations (a_g modes) of one or both parts of the complex.

In the infrared spectrum of ionic liquid-stabilized TTF·TCNQ nanoparticles, vibration bands for $BMIMBF_4$ or $DMIMBF_4$ are not observed. However, its presence and thus its stabilizing role have been evidenced by X-ray photoelectron spectroscopy. Indeed, the X-ray photoelectron spectrum shows boron and fluorine lines which can only be due to the ionic liquid. Moreover, the $S2p_{3/2}$ signal (163.6 eV) is in excellent agreement with that previously reported for TTF·TCNQ single crystals, i.e. 163.8 eV (Butler et al., 1974). In the spectrum of ionic liquid/oleic acid (OA)-stabilized TTF·TCNQ nanoparticles, vibration bands for OA are present, thus confirming that oleic acid forms a protecting layer around the nanoparticles. However, this stabilizing role is effective when oleic acid is at least introduced in a volume three times higher than that of the ionic liquid. Whatever the stabilizing agent, the infrared spectrum of nanoparticles recorded at room temperature in a KBr matrix (figure 7) is quite similar to that previously described for TTF·TCNQ processed as thin films (Wozniak et al., 1975; Benoit et al., 1976). Moreover, the infrared spectrum is weakly particle size-dependent. Peak positions and assignments are given in table 2.

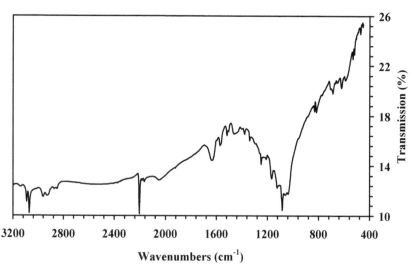

Fig. 7. Infrared spectrum at 298 K for TTF·TCNQ nanoparticles dispersed in KBr matrix (mean diameter: 35 nm; stabilizing agent: BMIMBF₄)

Assignment	νCH	νCH	νCN	νCN	νC=C	νC=C	–	–	δS–C–H	νC–S
1	3093	3073	2204	2182	1571	1518	1340	1170	1083	832/817
2	–	–	2207	2190	1552	1537	1355	1182	1096	797/786
3	–	–	very broad	–	1580	–	–	–	1080	798

Table 2. Infrared modes (cm^{-1}) and assignments for TTF·TCNQ nanoparticles (mean diameter: 35 nm; stabilizing agent: BMIMBF$_4$) and thin films. 1: our work; 2: Wozniak et al., 1975; 3: Benoit et al., 1976

Our spectrum clearly evidences C(sp^2)–H at 3093 and 3073 cm^{-1}. These modes are surprisingly not observed for TTF·TCNQ thin films deposited on NaCl or KBr crystals (Wozniak et al., 1975; Benoit et al., 1976). In our case, the characteristic nitrile doublet is located at 2204 and 2182 cm^{-1}. The more intense signal at 2204 cm^{-1} allows us to determine the amount of charge transfer from the TTF donor molecule to the TCNQ acceptor molecule. Using the linear correlation of the nitrile stretching mode for TCNQ as a function of the degree of charge transfer, we obtain a value of 0.56, in relatively good agreement with that for single crystals, *i.e.*, 0.59 (Chappell et al., 1981). Thus, the charge transfer in an assembly of nanoparticles is rather similar to that on a macroscopic single crystal. The position, the intensity and the width at half maximum for carbon-carbon double bond modes (in the 1500–1600 cm^{-1} range) are usually sample preparation and temperature dependent. For TTF·TCNQ nanoparticles, they are located at 1571 and 1518 cm^{-1}. Finally, the characteristic doublet for TTF (C–S stretch) is located at 832/817 cm^{-1} for us and at 797/786 cm^{-1} for TTF·TCNQ thin films. This large difference (about 30 cm^{-1}) can be an effect of the temperature (spectra recorded at room temperature in our case and at 10 K for TTF·TCNQ thin films, see Wozniak et al., 1975).

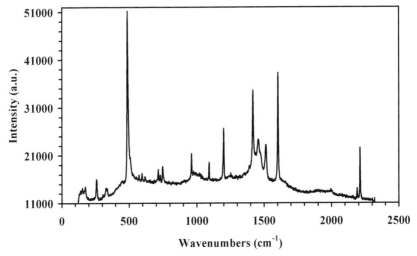

Fig. 8. Raman spectrum at 77 K for TTF·TCNQ as a nanoparticle film (mean diameter of individual particles: 35 nm; stabilizing agent: BMIMBF$_4$)

Raman spectroscopy is widely used to study the vibrational and structural properties of molecule-based conductors as single crystals. This form is the best suited for obtaining the best signal to noise ratio. However, this technique is also well suited for studying thin molecular layers (oriented, Langmuir-Blodgett, nanoparticle, or polymer films). The TTF·TCNQ nanoparticles are dispersed in diethyl ether and deposited on a glass slide. The solvent is then evaporated slowly. Raman spectra are obtained at 77 K using the 647 nm line of a Kr laser (power ~ 1.7×10^6 W cm^{-2}). The incident beam is focused onto the film through the ×100 microscope objective, giving a spot size of ~ 1 µm². The back-scattered light is collected through the same objective, dispersed and then imaged onto a CCD detector. Whatever the ionic liquid used as stabilizing agent, the Raman spectrum of the nanoparticle film (figure 8) is quite similar to that previously described for TTF·TCNQ single crystals

ν (cm^{-1})	Assignment	Symmetry
260 (263)	–	–
339 (333)	Ring deformation in TCNQ	a_g (ν_9)
489 (502)	C–S stretch and C–S–C bend	b_{3g} (ν_{47})
573 (572)	–	b_{2g} (ν_{29})
596 (600)	C(CN)$_2$ scissor	a_g (ν_8)
716 (714)	C–C ring stretch	a_g (ν_7)
748 (755)	C–S stretch	b_{2g} (ν_{28})
962 (962)	C–C ring stretch	a_g (ν_6)
1200 (1202)	C–C–H bend and C=C ring stretch in TCNQ	a_g (ν_5)
1418 (1423)	C=C stretch in TCNQ	a_g (ν_4)
1461 (1456)	C=C stretch centre and C=C stretch ring in TTF	–
1516 (1520)	C=C stretch centre and C=C stretch ring in TTF	a_g (ν_3)
1604 (1606)	C=C ring stretch in TCNQ	–
2210 (2224)	C≡N stretch	a_g (ν_2)

Table 3. Raman modes, assignments, and symmetry for TTF·TCNQ as a nanoparticle film (mean diameter of individual particles: 35 nm; stabilizing agent: BMIMBF$_4$) and as single crystals. In parentheses: values from Graja, 1997 or Kuzmany & Stolz, 1977

(Kuzmany & Stolz, 1977). An assignment of nearly all of the peaks can be found by comparing our results with the latter. In table 3, we compile the most important lines, their assignment, and their symmetry.

Except the nitrile stretching mode, all peak positions for TTF·TCNQ nanoparticles are very similar to those on single crystals (see values in parentheses on table 3). The C≡N stretching mode for TTF·TCNQ nanoparticles is located at 2210 cm⁻¹ in the Raman spectrum whereas it is located at 2204 cm⁻¹ in the infrared spectrum (figure 7 and table 2). This difference is still relatively low given the uncertainties on the positions of signals ($\pm$ 4 cm⁻¹) for both spectral techniques. The following discussion will be mainly based on the totally symmetric a_g modes. These modes (10 for the TCNQ molecule and 7 for the TTF molecule) have been intensively studied for neutral (TTF⁰, TCNQ⁰), anionic (TCNQ⁻), or cationic (TTF⁺) species (Graja, 1997). The conduction electrons in TTF·TCNQ are significantly coupled to almost all intramolecular vibration (a_g) modes. The more significant ones (from v_2 to v_6) have been used to investigate the charge transfer in TTF·TCNQ single crystals (Kuzmany & Stolz, 1977). Table 4 gathers a_g v_2 to v_6 modes (column two) for TTF·TCNQ nanoparticles, for neutral TCNQ (column three) and for the TCNQ⁻ anion (column four). Comparing column two with columns three and four in table 4, evidences that the Raman lines of TTF·TCNQ are generally within the limits of neutral and completely charged TCNQ.

	TTF·TCNQ nanoparticles	TCNQ⁰	TCNQ⁻	ρ
v_2	2210	2230	2192	−0.53
v_3	1604	1602	1613	−0.18
v_4	1418	1453	1389	−0.55
v_5	1200	1207	1195	−0.58
v_6	962	932	960	−1.07

Table 4. A_g modes (in cm⁻¹) for TTF·TCNQ nanoparticles, neutral TCNQ and radical anion, and evaluation of the charge transfer in TTF·TCNQ as nanoparticles

According to Graja, the charge (ρ) borne by the TCNQ molecule can be calculated as follows:

$$\rho = \frac{v - v_0}{v_0 - v_{-I}} \quad (5)$$

where v is the vibrational frequency for TTF·TCNQ, v_0 the vibrational frequency for neutral TCNQ, and v_{-I} the vibrational frequency for TCNQ⁻. Values of ρ obtained from equation (5) are given in table 4. Kuzmany & Stolz admit that the negative charge borne by the TCNQ molecule in TTF·TCNQ is the average of all charges as compiled in table 4 (column five). In our case, the average charge is found to be –0.58, in excellent agreement with that for single crystals, *i.e.*, –0.59 (Chappell et al., 1981).

5. Conclusion

In this chapter, we have shown that the famous organic charge transfer-based conductor TTF·TCNQ can be processed as roughly spherical nanoparticles, while this compound has a natural tendency to grow as needles. The growth as nanospheres is controlled by the use of an ionic liquid introduced together with a conventional solvent. Under certain conditions, well dispersed nanoparticles exhibiting sizes lower than 50 nm can be obtained. Optical and vibrational studies have been performed on either TTF·TCNQ dispersed in an infrared transparent matrix, or on a nanoparticle film, or dispersed in solution. Optical and vibrational signatures for TTF·TCNQ as nanoparticles are in good agreement with those on single crystals, on powders exhibiting micrometer-sized grains, or on thin films. We do expect transport properties similar as the two latter. Preliminary conductivity measurements on a nanoparticle film (mean diameter of individual nanoparticles: 4 nm or 35 nm) have been performed using the four probe technique. TTF·TCNQ nanoparticle films exhibit a semiconducting behavior, which is not surprising for nanopowdered materials (room-temperature conductivity: 10–20 S cm^{-1}, activation energy: 15–40 meV). We are currently working on the use of dielectric spectroscopy to characterize an assembly of TTF·TCNQ nanoparticles. We are also developing the use of Raman spectroscopy (at low temperatures) to investigate the charge transfer in nanoparicles of organic or metallo-organic superconducting phases.

6. Acknowledgment

The authors appreciate the contribution and collaboration of the following persons: C. Faulmann (LCC–Toulouse, France); H. Hahioui (University of Rabat, Morocco); C. Routaboul (LCC–Toulouse and University Paul Sabatier, France); F. Courtade and J.-M. Desmarres (CNES–Toulouse, France); O. Vendier (Thales Alenia Space–Toulouse, France); J. Fraxedas (CIN2–Barcelona, Spain).

7. References

The order of reference is that of citation in the text.
Ferraris, J.; Cowan, D. O.; Walatka, Jr., V. & Perlstein, J. H. (1973). Electron transfer in a new highly conducting Donor–Acceptor complex. *Journal of the American Chemical Society,* Vol. 95, No. 3, (February 1973), pp. 948-949.
Kistenmacher, T. J.; Phillips, T. E. & Cowan D. O. (1974). The crystal structure of the 1:1 radical cation–radical anion salt of 2,2'-bis-1,3-dithiole (TTF) and 7,7,8,8-tetracyanoquinodimethane (TCNQ). *Acta Crystallographica B,* Vol. 30, No. 3, (March 1974), pp. 763-768.
Ikemoto, I.; Sugano, T. & Kuroda H. (1977). Evaluation of the charge transfer in tetrathiofulvalene-tetracyanoquinodimethane (TTF-TCNQ) and related complexes

from X-ray photoelectron spectroscopy. *Chemical Physics Letters*, Vol. 49, No. 1, (July 1977), pp. 45-48.

Coppens, P. (1975). Direct evaluation of the charge transfer in tetrathiafulvalene-tetracyanoquinodimethane (TTF-TCNQ) complex at 100 K by numerical integration of X-ray diffraction amplitudes. *Physical Review Letters*, Vol. 35, No. 2, (July 1975), pp. 98-100.

Chappell, J. S.; Bloch, A. N.; Bryden, W. A.; Maxfield, M.; Poehler, T. O. & Cowan, D. O. (1981). Degree of charge transfer in organic conductors by infrared absorption spectroscopy. *Journal of the American Chemical Society*, Vol. 103, No. 9, (May 1981), pp. 2442-2443.

Fraxedas, J.; Molas, S.; Figueras, A.; Jiménez, I.; Gago, R.; Auban-Senzier, P. & Goffman, M. (2002). Thin films of molecular metals: TTF-TCNQ. *Journal of Solid State Chemistry*, Vol. 168, No. 2, (November 2002), pp. 384-389.

Yan, H.; Li, S.; Yan, C.; Chen, Q. & Wan, L. (2009). Absorption of TTF, TCNQ and TTF-TCNQ on Au(111): An *in situ* ECSTM study. *Science in China Series B: Chemistry*, Vol. 52, No. 5, (May 2009), pp. 559-565.

Savy, J.-P.; de Caro, D.; Faulmann, C.; Valade, L.; Almeida, M.; Koike, T.; Fujiwara, H.; Sugimoto, T.; Fraxedas, J.; Ondarçuhu, T. & Pasquier, C. (2007). Nanowires of molecule-based charge-transfer salts. *New Journal of Chemistry*, Vol. 31, No. 4, (April 2007), pp. 519-527.

Gutel, T.; Garcia-Antòn, J.; Pelzer, K.; Philippot, K.; Santini, C. C.; Chauvin, Y.; Chaudret, B. & Basset, J. M. (2007). Influence of the self-organization of ionic liquids on the size of ruthenium nanoparticles: effect of the temperature and stirring. *Journal of Materials Chemistry*, Vol. 17, No. 31, (August 2007), pp. 3290-3292.

Grant, P. M.; Greene, R. L.; Wrighton, G. C. & Castro, G. (1973). Temperature dependence of the near-infrared optical properties of tetrathiofulvalinium-tetracyanoquinodimethane (TTF-TCNQ). *Physical Review Letters*, Vol. 31, No. 21, (November 1973), pp. 1311-1314.

Bright, A. A.; Garito, A. F. & Heeger, A. J. (1974). Optical conductivity studies in a one-dimensional organic metal: tetrathiofulvalene tetracyanoquinodimethane (TTF)(TCNQ). *Physical Review B*, Vol. 10, No. 4, (August 1974), pp. 1328-1342.

Graja, A. (1997). *Spectroscopy of materials for molecular electronics* (first edition), Scientific Publishers OWN and Polish Academy of Sciences, 83-85481-23-0, Poznań

Butler, M. A.; Ferraris, J. P.; Bloch, A. N. & Cowan, D. O. (1974). Electron transfer in TTF-TCNQ and related compounds. *Chemical Physics Letters*, Vol. 24, No. 4, (February 1974), pp. 600-602.

Wozniak, W. T.; Depasquali, G.; Klein, M. V.; Sweany, R. L. & Brown, T. L. (1975). Vibrational spectra of TTF-TCNQ: evidence of TTF^0 and $TCNQ^0$ in thin films. *Chemical Physics Letters*, Vol. 33, No. 1, (May 1975), pp. 33-36.

Benoit, C.; Galtier, M.; Montaner, A.; Deumie, J.; Robert, H. & Fabre, J.-M. (1976). Temperature dependence of infrared spectra of TTF(TCNQ) thin films. *Solid State Communications*, Vol. 20, No. 3, (October 1976), pp. 257-259.

Kuzmany, H. & Stolz, H. J. (1977). Raman scattering of TTF-TCNQ and related compounds. *Journal of Physics C: Solid State Physics,* Vol. 10, No. 12, (June 1977), pp. 2241-2252.

6

Probing Metal/Organic Interfaces Using Doubly-Resonant Sum Frequency Generation Vibrational Spectroscopy

Takayuki Miyamae
Nanosystem Research Institute,
National Institute of Advanced Industrial Science and Technology (AIST),
Japan

1. Introduction

Recent advances in the performance of organic semiconductors such as organic light emitting diodes (OLEDs), organic field-effect transistors (OFETs), and organic solar cells have been dramatic. In particular for OLEDs, stability and durability have improved to levels that warrant their application in everyday life. In organic devices, the charge carriers, both electrons and holes, often have to be injected through organic/electrode interfaces. Therefore, an understanding of the interaction between metal electrode and organic molecules is quite important, because the electronic properties of metal/organic interface directly affect the performance of the OLEDs. (Tang & Slyke, 1987; Salaneck et al., 2002) But these are not well examined, especially the physical and chemical properties of buried organic/meal interfaces have been not well investigated. In order to understand the mechanisms that control the electron energetic levels of organic/metal interfaces, the determination of the energy barriers between the Fermi level of the metal and the HOMO and LUMO levels of organic materials across the interfaces has been main challenges to surface and interface studies of organic thin films. For understanding the occupied states, ultraviolet photoelectron spectroscopy (UPS) is one of the powerful techniques for studying the valence electronic structure of material. The electron injection barrier at the metal/organic interface is significantly altered by the interfacial dipole layer, by which the vacuum level at the organic layer is shifted relative to that at the metal layer. (Ishii et al., 1999) The dipole layer has been studied for organic/metal interfaces by using photoelectron spectroscopic technique. Ishii et al. proposed various origins of the dipole layer at the organic/metal interface: (1) charge transfer, (2) mirror force, (3) pushback effect due to the surface rearrangement, (4) chemical interaction, (5) interface state, and (6) permanent dipole of the adsorbate. (Ishii et al., 1999)

In contrast, the band gaps of the organic materials and the energy of LUMO are often determined by the optical absorption measurements of the *"bulk"*. Because of the surface confinement effect and the interaction between organic molecules and metal, the molecular conformation and the band gap at the buried interface are expected to be different from those in the bulk. Although the optical band gap obtained from the absorption spectroscopy is different from the corresponding charge transfer gap due to the exciton absorption to form Frenkel-type exciton, the optical band gaps at buried interfaces are still useful energy parameter to discuss the charge injection and overall efficiency of the OLED. However, it has been a great challenge to measure the buried interfacial electronic states because of a lack of a suitable probing technique. Traditional surface science techniques based on ultrahigh vacuum are not applicable to a buried interface, and absorption and emission spectroscopy do not have the necessary surface sensitivity.

Nonlinear optical spectroscopy is one of the powerful techniques for the characterization of these issues due to its high interface sensitivity. Second harmonic generation (SHG) and sum frequency generation (SFG) have been used to investigate the molecular orientation of the materials at the interface. (Shen, 1984) IR-visible SFG spectroscopy has made it possible to study the vibrational spectra of surface or interfacial species. SFG is a surface-sensitive tool because its second-order nonlinear optical process is allowed only in non-centrosymmetric media under the electric dipole approximation. Recently this technique has been applied to interfaces that include organic materials, allowing the interfacial structures to be elucidated. In our SFG studies of polymer/water, (Miyamae et al., 2007) polymer/oxide, (Miyamae & Nozoye, 2004) and air/liquid interfaces, (Miyamae et al., 2008; Iwahashi et al., 2008) we have found that the molecular structures, such as the orientation and orientation distribution of different chemical groups, are not the same at the different interfaces.

IR-visible SFG spectroscopy has been traditionally carried out by using the frequency-fixed visible and tunable IR beams to obtain a surface vibrational spectrum, which identifies the surface chemical species. Recently, a new technique for vibrational SFG spectroscopy by tuning the incident visible and IR frequencies has attracted much attention. (Raschke et al., 2002) When the photon energy of the SFG coincides with electronic transition energies of interface species, the output SFG intensity is drastically enhanced when the IR light is resonant with the vibrational state and the output SFG light is resonant with the interfacial electronic transitions. Such enhanced SFG process is called doubly-resonant (DR) SFG. With the capability of tuning both the incident IR and visible frequencies, doubly-resonant SFG spectroscopy becomes a powerful multi-dimensional technique for studying the interface electronic states coupled to a specific vibrational mode. The SFG electronic excitation profiles, which can be obtained by measuring the visible probe frequency dependence of the vibrational SFG band strengths, allow deduction of coupling electronic transitions and vibrational modes at the interface. (Miyamae et al., 2009)

In addition, there are several advantages of the doubly-resonant SFG technique. The signal enhancement is expected only for species that have an electronic absorption at the photon energy of the SFG. Therefore, doubly-resonant effect offers a kind of molecular selectivity to SFG. Moreover, an electronic excitation spectrum of the interface species for each vibrational band can be obtained. Thus the SFG excitation profiles are useful to investigate mixed interface layers where several chemical species coexist and show complex vibrational spectra, because vibrational bands can be classified with reference to corresponding electronic spectra. In addition, the measurement of SFG excitation profiles may be an effective way to obtain electronic spectra of the molecules, especially at interface on opaque substrates where electronic absorption spectrum measurement is difficult. Furthermore, it can be possible to measure the interface of the OLED materials that show very strong photoluminescence in visible region, since the output SFG emerges at the Anti-Stokes side of the excitation wavelength.

This chapter is constructed as follows: in Section 2, we describe the theoretical backgrounds of the DR-SFG processes. Section 3 describes the experimental setup for the DR-SFG measurements and the sample preparation. Section 4 illustrates the DR-SFG study of the electronic and vibrational properties of the OLED interfaces. The chemical structures of all the organic compounds involved in this chapter are sketched in Figure 1.

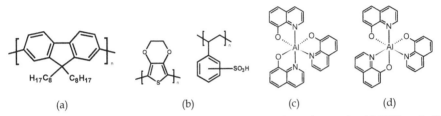

(a)   (b)   (c)   (d)

Fig. 1. Chemical structure of the various organic compounds under study: (a) PFO, poly(9,9-dioctyl-fluorene); (b) PEDOT:PSS; (c) meridional Alq$_3$, tris(8-hydroxyquinoline) aluminum; (d) facial Alq$_3$.

2. Theoretical background of the doubly-resonant sum frequency generation.

In the electric dipole approximation, SFG is forbidden in centrosymmetric materials, but not at their interfaces, where the inversion symmetry of the bulk is broken. For such an air-metal interface, the SFG intensity reflected from the surface is given by

$$I(\omega_{SF}) \propto \left| \chi_{eff}^{(2)} : E(\omega_{IR})E(\omega_{VIS}) \right|^2 ,$$
(1)

where $\chi^{(2)}_{eff}$ is the effective second-order nonlinear susceptibility tensor and $E(\omega_{IR})$ and $E(\omega_{VIS})$ are the input fields. The second-order nonlinear susceptibility contains nonresonant, singly resonant, and doubly resonant contributions. The latter dominates strongly if the vibrations (ω_l) and electronic transitions (ω_{eg}) probed by ω_{IR} and ω_{SFG} are coupled. Both IR–visible (vibrational transition followed by an electronic transition) and visible–IR processes (electronic transition followed by a vibrational transition) contribute to doubly-resonant SFG.(Raschke et al., 2002; Hayashi et al., 2002) However, the IR–visible sequence is expected to dominate, as a result of the quicker relaxation of the electronic excitation compared to the vibrational one. Assuming harmonic potential surfaces for the electronic states and the Born-Oppenheimer and Condon approximations, the doubly–resonant $\chi^{(2)}_{ijk}$ can be described as

$$\chi_{ijk}^{(2)} = -\frac{N}{\hbar^2} \left\langle \mu_{eg}^i \mu_{ge}^j \frac{\partial \mu_{gg}^k}{\partial q_l} \frac{\sqrt{S_l} e^{-S_l}}{\omega_{IR} - \omega_l + i\Gamma_l} \right.$$

$$\left. \times \sum_{n=0}^{\infty} \frac{S_l^n}{n!} \left[\frac{1}{\omega_s - n\omega_l - \omega_{eg} + i\Gamma_{en,g0}} - \frac{1}{\omega_s - (n+1)\omega_l - \omega_{eg} + i\Gamma_{en+1,g0}} \right] \right\rangle + \chi_{NR,ijk}^{(2)} ,$$
(2)

where N is the surface molecular density, μ_{eg}^i represents the i component of electronic transition moment, q_l is the normal coordinate, S_l is a dimensionless coupling constant known as the Huang-Rhys factor, n labels the vibrational state, g and e label the ground and excited electronic states, respectively, ω_S is the SFG frequency, ω_l and ω_{eg} are the resonant vibrational and electronic frequencies, respectively, Γ_l and $\Gamma_{en,g0}$ are the damping constants, the angular brackets indicate an average over molecular orientations, and $\chi^{(2)}_{NR,ijk}$ describes the non-resonant contributions. S_l is related to the shift d_l of the harmonic potential of the vibration in the excited electronic level by

$$S_l = \frac{1}{2\hbar}\omega_l d_l^2 . \tag{3}$$

As shown in Fig. 2, DR-SFG occurs thus for $\omega_{IR} = \omega_l$ and for several visible frequencies, when ω_S matches an allowed vibronic transition to the excited electronic level. The intensity of each vibronic resonance depends on the Frank-Condon overlap integrals of the vibrational levels involved in the transition. Since the initial and final vibrational states, respectively of the visible and SFG transitions, always differ, the vibration and the electronic transition must therefore be coupled ($d_l \neq 0$) to have a non-zero transition probability for the global DR-SFG process. Thus the DR-SFG spectrum allows for the determination of the coupling strength and characteristics.

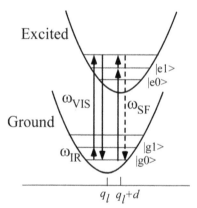

Fig. 2. Energy diagram for DR-SFG. The vibration harmonic potentials are shifted between the ground and excited states. The resonant transitions involved in the DR IR–visible (and visible–IR) sum-frequency generation processes are shown schematically.

Equation (2) includes all the vibronic transitions series. However, in general, the visible–IR SFG is much weaker and not detected because of the very fast relaxation of the electronic excitation. For example, the dephasing times of vibronic transitions are in the femtosecond region for Alq_3. Therefore, by assuming $\Gamma_{en,g0} >> \Gamma_{e0,g0}$, the nonzero vibronic transitions can be neglected.

It is worth pointing out that a significantly larger $\Gamma_{en,g0}$ also suppresses the aforementioned visible–IR SFG, which starts with an electronic transition followed by a vibrational transition.

To analyze the spectra, we note that with the visible input frequency ω_{vis} fixed, Eq. (2) can be approximated by the form

$$\chi_{ijk}^{(2)} \propto \sum_l \frac{A_l}{\omega_{IR} - \omega_l + i\Gamma_l} + \chi_{NR}^{(2)} e^{i\xi} , \tag{4}$$

where A_l, ω_l, Γ_l are the peak amplitude describing the electronic resonance, the resonant vibrational frequencies, and the damping constants, respectively. $\chi^{(2)}_{NR,ijk}$ and ξ describes the non-resonant contributions and the phase difference between resonant and non-resonant term, respectively. We use Eq. (4) to fit all the measured spectra with ω_l, Γ_l, A_l, ξ, and $\chi^{(2)}_{NR}$ as adjustable parameters.

3 Experimental methods and materials

3.1 Doubly-resonant sum frequency generation
Figure 3 schematically depicts the SFG experimental setup for the doubly-resonant SFG measurements. (Miyamae et al., 2009 & 2011) Tunable IR and visible laser beams were generated by two optical parametric generators/amplifiers (OPG/OPA, Ekspla, PG401VIR/DFG) pumped by a mode-locked Nd:YAG laser at 1064 nm (Ekspla, PL-2143D, 25ps, 10Hz). The IR beam, tunable from 1000cm⁻¹ to 4300 cm⁻¹, was produced by difference frequency mixing of the 1064 nm beam with the output of a LiB₃O₅ (LBO) crystal mounted in OPG/OPA, which is pumped by the 355 nm beam. The visible beam, tunable from 420 to 640 nm, was generated in a LBO crystal mounted in OPG/OPA pumped by the 355 nm beam. The visible and IR beams were overlapped at sample surface with the incidence angles of 70° and 50°, respectively. The spectral resolution of the tunable visible beam was about 8 cm⁻¹, and its frequency was calibrated with the Hg lines. The spectral resolution of the IR beam was 6 cm⁻¹, and the IR frequency was calibrated with the absorption lines of polystyrene standard. In order to minimize the irradiation damage, both tunable infrared and visible beams were defocused. The focus size of the infrared and visible beams were >1 mm and >3 mm, respectively. Further, in order to avoid photo-irradiated damage, the fluence of the visible beam was kept below 100 μJ per pulse. The absence of the damage effect was confirmed by repeated SFG measurements. In order to eliminate the scattered visible light and the photoluminescent light from the samples, the sum-frequency output signal in the reflected direction was filtered with short-wave-pass filters (Asahi Spectra Co. Ltd.), prism monochromator (PF-200, Bunkoukeiki Co., Ltd.), and grating monochromator

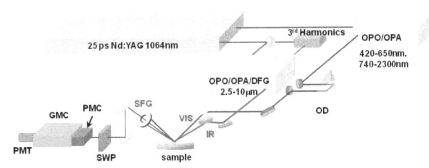

Fig. 3. The schematic arrangement of the SFG spectrometer for the DR-SFG. OD; Optical delay, SWP; short wave pass filters, PMC; prism monochromator, GMC; grating monochromator, PMT; photomultiplier tube.

(Oriel MS257). Then the SFG signal was detected by a photomultiplier tube (Hamamatsu R649). The signal was averaged over 300 pulses by a gated integrator for every data point taken at a 3 cm^{-1} interval and was stored in a personal computer. In the frequency region between 2000 and 1300 cm^{-1}, significant portion of the infrared beam is absorbed by water vapor in the optical path. The effect was minimized by purging the optical path of the IR beam and the sample stage by dry nitrogen gas. Each SFG spectrum was normalized to the visible and IR power to compensate the laser intensity fluctuations.

3.2 Sample preparations for the SFG measurements
3.2.1 Preparation of the PFO samples
The samples of the PFO endcapped with dimethylphenyl (MW=48.8K) were obtained from American Dye Source and used as received. Thin film of the PEDOT:PSS solution (CREVIOS Al4083) was spin-coated at 3000 rpm on the top of the fused quartz or CaF$_2$ substrates and then baked at 150 °C for 60 min in an oven with N$_2$ flow. PFO layer was then spin-coated at 3000 rpm from 1% w/v toluene solution on top of the PEDOT:PSS layer. Then the films were evacuated to eliminate residual solvent.

For the measurements of the buried Al/LiF/PFO interface, the LiF and the Al were directly deposited on the spin-coated PFO/PEDOT:PSS onto CaF$_2$ substrates in a vacuum chamber. The thicknesses of the LiF and Al are about 1 and 40 nm, respectively.

3.2.2 Preparation of the Alq$_3$ samples
For the preparation of the Alq$_3$/Al samples, the Al substrates were prepared by vacuum evaporation of Al on Si substrates by using a tungsten filament. Here, the notation of A/B indicates a system prepared by depositing A on B. Deposition of Alq$_3$ was performed using a quartz crucible coiled with a tungsten wire heater. Alq$_3$ were deposited on them under dark condition in a vacuum chamber. In order to avoid the influence of the air, 50 nm thick CaF$_2$ was deposited after the deposition of the 2 nm thick Alq$_3$. For the observation of the Al/Alq$_3$ interfaces by SFG, the Alq$_3$ films were directly deposited on CaF$_2$ substrates, and thick Al layers were then deposited under dark condition. For Al/LiF/Alq$_3$, thin LiF film of 1 nm thick was deposited on the Alq$_3$ from a tungsten basket. Then 50 nm thick Al layer was deposited on the LiF layer. After the deposition, the SFG measurements were subsequently performed in ambient conditions.

4. DR-SFG study of organic/metal Interfaces

4.1 DR-SFG study of poly(9,9-dioctylfluorene) surfaces and Al/LiF interfaces
Polymer LEDs are one of the most promising applications given the current high interest in developing ultra thin computer monitors and television sets, i.e., flat-panel displays. The research on polymers in LEDs has undergone a rapid expansion beginning in 1990, when results on a light-emitting diode with poly(p-phenylenevinylene) (PPV), as the emitting layer was published. (Burroughes et al., 1990) Recently, one particular classes of conjugated polymers, the poly(9,9-dioctylfluorene) (PFO, chemical structure is shown in Fig. 1) and fluorene-arylene copolymers have been studied intensively because of its applications in the LEDs due to their highly efficient blue photoluminescence. (Mallavia et al., 2005) Although the bulk electronic and optical properties of PFO have been studied extensively by UV-visible absorption, Raman, and photoluminescence spectroscopy,(Ariu et al., 2000; Montilla

& Mallavia, 2007) the optical and electronic properties of PFO at the buried electrode interface remain unexplored. Because the electronic properties of electrode/organic interface affect the performance of the organic LEDs, understanding of the interaction between the electrode and organic molecules and the electronic structures at the buried interfaces are quite important. In an organic device, the charge carriers have to be injected through polymer/electrode interfaces. Therefore, the band gaps of conjugated polymers at the buried interface are important energy parameters to discuss the charge injection and overall efficiency of the organic devices. In this study, we present the surface and the buried interfacial vibrational and electronic structure of the PFO using SFG. (Miyamae et al., 2010) Figure 4 shows the SFG vibrational spectra from the PFO/PEDOT:PSS/quartz surface with various visible wavelengths in a SSP configuration (S-, S-, and P-polarized for SFG, visible, and IR, respectively). SFG susceptibility of the quartz is expected to be approximately constant in the visible region because it is transparent to visible light. A strong vibrational band was observed at 1610 cm-1 in all spectra, and the band intensity increased when the visible probe wavelength was changed from 550 to 435 nm. The vibrational band at 1610 cm-1 is assigned to the C=C symmetric stretching of the fluorene rings located at the backbone of PFO. The electronic resonance enhancement of the SFG spectra is observed when the visible wavelength is near 435 nm, which produces a SFG wavelength near 407 nm with an IR beam of 1610 cm-1. It should be noted that the PFO surface does not change during the SFG measurements. It has been reported that the oxidative defect are formed during the operation or UV light irradiation. (List & Guentner, 2002; Gong et al., 2003) The UV-visible and IR absorption signature allow an identification of the defects as ketone groups attached to the 9-position of the fluorene unit; thus, the fluorene unit becomes a 9-fluorenone due to oxidative degradation. The SFG spectra clearly shows that the absence of the oxidative peak derived from the fluorenone, indicating that the PFO surface is not oxidized during the SFG measurements.

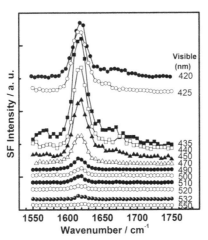

Fig. 4. SFG vibrational spectra of PFO surface with various incident visible wavelengths. Spectra are offset for clarity

Next, we show in Figure 5 the SFG spectra from the buried Al/LiF/PFO interface with various visible wavelengths in SSP polarization combination. In this project, the LiF and the Al layers were directly deposited on the spin-coated PFO/PEDOT:PSS onto CaF$_2$ substrate. Since the deposited Al layer can act as superior gas barrier, the extent of the oxidation of the Al interfaces is expected to be much reduced. The vibrational band at 1610 cm⁻¹ is still observed in all SFG spectra. As discussed later, the 1610 cm⁻¹ peak is not derived from the buried PFO/PEDOT:PSS interface due to the less orientational order at the polymer/polymer interface. The peak position of the band does not change between the PFO surface and the PFO interface, indicating that the PFO is not degradated by the Al deposition. In the case of the DR-SFG spectra of the Al/LiF/PFO interface, it should be noted that the SFG peak shows different shape from that of the air/PFO interface. This difference is caused by the different interference phenomena with the SFG non-resonant contribution arising from the Al substrate.

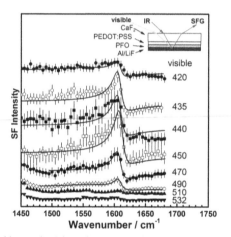

Fig. 5. The SFG spectra of buried Al/LiF/PFO interface with various incident visible wavelengths. The experimental data and the fitted curves using eq. 4 are represented by dots and solid lines, respectively. Experimental SFG setup for probing the buried interface is shown in the inset. Spectra are offset for clarity.

Because of the electronic transitin of the PFO, the refractive index of PFO in the investigated region is wavelength-dependent. (Campoy-Quiles et al., 2005) Therefore, the Fresnel factors need to be considered to obtain the actual dispersion relation of the second-order nonlinear susceptibility. In the electric-dipole approximation, the effective second-order nonlinear susceptibility tensor for the SSP polarization can be written as

$$\chi_{eff}^{(2)} = L_{YY}(\omega_{SF})L_{YY}(\omega_{VIS})L_{ZZ}(\omega_{IR})\sin\beta_{IR}\chi_{yyz}^{(2)}, \tag{5}$$

where $L_{YY}(\omega)$ and $L_{ZZ}(\omega)$ are the Fresnel coefficients at frequency ω; β_{IR} is the reflection angle of the IR frequency; and χ_{yyz} is the nonvanishing yyz component of the second-order nonlinear susceptibility in the laboratory coordinate. In the SSP polarization, the Fresnel factor can be written as

$$F_{yyz} = |L_{YY}(\omega_{SF})L_{YY}(\omega_{VIS})L_{ZZ}(\omega_{IR})\sin\beta_{IR}|, \tag{6}$$

The refractive indices for PFO and metallic aluminum reported by Campoy-Quiles et al. (Campoy-Quiles et al., 2005) and the literature (Parik, 1985) were used for the evaluation of the Fresnel factors.

In general, there are two types of processes in IR-visible SFG, as mentioned in the theoretical section. The first type starts with an electronic transition followed by a vibrational transition (VIS-IR SFG), and the second type begins with a vibrational transition followed by an electronic transition (IR-VIS SFG). (Hayashi et al., 2002) Because the electronic relaxation times are generally much shorter than the vibrational relaxation times, the contribution of the VIS-IR SFG is generally negligible. If the VIS-IR SFG occurs, increase of the non-resonant background is expected due to the ultrafast dephasing dynamics of the S_1 state. (Wu et al., 2009) Therefore, only the IR-VIS SFG will be considered in the following analysis.

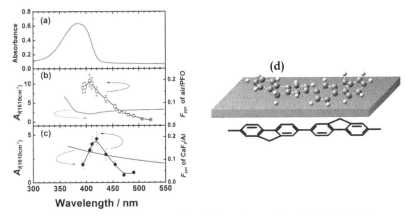

Fig. 6. (a) Optical absorption spectrum of the PFO spin cast film. (b) The SFG electronic excitation profile of air/PFO interface and the Fresnel factor F_{yyz} (continuous blue curve) at air/PFO interface, and (c) the SFG electronic excitation profile of Al/LiF/PFO interface and the F_{yyz} at CaF$_2$/Al interface (red curve). Solid black lines are guide to the eyes. (d) Proposed model for the planar configuration at the interface.

The curves b and c in Figure 6 show the changes in the A_l of the 1610 cm^{-1} peak deduced from the fitting of the DR-SFG spectra in Figs. 4 and 5 as a function of the photon energies of the SFG. Figure 6a shows the optical absorption spectrum of the PFO spin cast film. The broad optical absorption band originates from inhomogeneously broadened $\pi \rightarrow \pi^*$ transitions of the glassy PFO. (Cadby et al., 2000) As shown in Fig. 6, the SFG electronic excitation spectrum obtained from the air/PFO interface is slightly red-shifted with respected to an optical absorption maximum of PFO film. The SFG excitation spectrum at the Al/LiF/PFO interface is also plotted in Fig. 6c, and it is further red-shifted with respected to that of the PFO surface. It should be noted that these red-shifts are not caused by the visible variations of the Fresnel factors. As shown in Fig. 6, the wavelength variations of the F_{yyz} both air/PFO and CaF$_2$/Al interfaces do not explain the SFG electronic excitation profiles of the air/PFO and the buried interface. Thus we conclude that the changes of the Fresnel factors do not much affect on the spectral shape of the SFG excitation profile of the 1610 cm^{-1} peak.

We attribute that the origin of the red-shifts of the SFG electronic excitation spectra from the air/PFO and Al/LiF/PFO interfaces are due to the stress-induced surface confinement effects of the polymer backbone, as in the case of the MEH-PPV interfaces. (Li et al., 2008) In general, the optical band gap of a conjugated polymer is closely related to the conjugation length. Conjugated polymer chains consist of a series of connected segments, each of which has a different extent of π-electron delocalization. Although the extent of the conjugation is limited by the twists in the polymer backbone, the longer the segment is, the smaller the optical band gap of the conjugated polymers due to the increasing average effective conjugation length. The restriction of the torsion angle between adjacent segments at the air/polymer and the Al/LiF/polymer interfaces produce a longer conjugation length. Similar red-shift of the electronic excitation profile at the interface has been also reported by the SFG measurements for the MEH-PPV interfaces. (Li et al., 2008)

To gain further structural information at the PFO surface, we performed the SFG measurements in the CH stretching region. The SSP SFG spectrum of the PFO surface taken at the visible wavelength of 532 nm exhibits peaks derived from the aliphatic hydrocarbon peaks that originate from the side chain of PFO, as shown in Fig. 7a. The tiny SFG signals derived from the CH stretching mode of PFO ring are observed around 3050cm-1 at the visible excitation wavelength of 532 nm. We also tried to measure the SPS and PPP polarization combinations, however, the intensities of the peak around 3050 cm-1 in SPS and PPP were much weaker than those of SSP spectra. One may think that the PFO surface must be covered with the side chain of the PFO because the peaks derived from the aliphatic hydrocarbon peaks are clearly detected. However, this 3050 cm-1 peak is clearly observed when the visible excitation wavelength is changed to shorter wavelength, as shown in Fig. 7. This result indicates that the planes of PFO rings are nearly parallel to the surface plane at the air/PFO interface, and the PFO surface is not fully covered with the aliphatic side chains. If the molecules take a planar orientation at the surface, it is hard to detect in SFG without the electronic resonance. It is well known that planar ordering of the polymer chains parallel to the interfaces due to the confinement effects occurs and this creates highly anisotropic optical properties to the film surface. (Kawana et al., 2002; Knaapila et al., 2003) The planar ordering of the polymer chains at the surface is in good agreement with the X-ray diffraction studies and the optical investigation by emission spectroscopy.(Cheum et al., 2009)

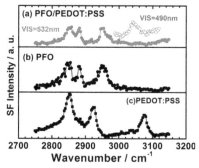

Fig. 7. SFG spectra of (a) PFO on quartz substrate taken by the 532nm (●) and 450nm (○), (b) PFO/PEDOT:PSS on quartz, and (c) PEDOT:PSS on quartz.

It is well known that the bulk solid PFO film exhibits complex phase behavior. Disordered PFO forms the glassy phase where the polymer backbones do not form a particular conformation with long-range order. In contrast, PFO in the so-called β-phase is extended conformation of PFO chains and possesses lower energy, due to the backbone planarization. (Grell et al., 1999) Single molecule spectroscopy demonstrates that the β-phase features of PFO are the results of stress-induced backbone planarization of polymer chain. (Becker &

Lupton, 2009) The optical absorption spectra of β-phase PFO exhibits the characteristic shoulder absorption around 430 nm in comparison with the glassy PFO. The SFG electronic profiles of the PFO interfaces have maximum around 410 – 420nm, and this peak position is close to the shoulder absorption of the β-phase PFO, rather than that of the glassy PFO. Because of the restriction of the torsion angle between adjacent segments, the conformation of the polymer backbone is limited at the polymer interfaces. As a result, the effective conjugation length at the interface is increased. Thus we conclude that the optical band gaps at the Al/LiF/PFO interfaces become smaller than that of the bulk glassy PFO, due to the stress-induced chain planarization at the air/PFO and the Al/LiF/PFO interfaces. The appearance of β-phase in the PFO films can be affected by many experimental treatments, such as, thermal annealing, organic solvent vapor exposure, mechanical stress, chemical modification of PFO chains, and so on. (Zhu et al., 2008) In addition to these treatments, our findings indicate that the interface confinement effect is also affected to induce the planar orientation of the polymer chains. Proposed planar orientation of the PFO chains at the interfaces is shown in Fig. 6 (d).

Figure 7 also indicates that the SFG signals are not derived from the buried PFO/PEDOT:PSS interface. To endure the observed SFG is truly generated at the air/PFO interface without a contribution from the buried PFO/PEDOT:PSS interface, we compare the SFG spectra for the PFO coated on PEDOT:PSS and the PEDOT:PSS films in the CH stretch region. The SFG spectrum of the PEDOT:PSS exhibits three vibrational resonances from CH stretches at 2855, 2927, and 3074 cm^{-1}. (Silva & Miranda, 2009) With an additional PFO layer, the SFG spectrum changes both in intensity and shapes. The SFG spectra of the PFO on PEDOT:PSS layer is very close to that of the PFO. Consequently, the topmost molecular orientation of the PEDOT:PSS does not affect the surface molecular orientation of the PFO. From this observation, we conclude that only the topmost layer has a net orientational order and contributes to the SFG spectra while the interface between PFO and PEDOT:PSS does not have contributions to the SFG spectra due to less orientational order that is inactive for SFG.

4.2 DR-SFG study of tris-(8-hydroxy-quinoline) aluminium (Alq$_3$)/Al interfaces
4.2.1 DR-SFG study of Alq$_3$ films: Peak assignments and the thickness dependences

In this section, DR-SFG was applied to detect the interfacial vibrational and electronic states of tris(8-hydroxyquinoline) aluminum (Alq$_3$)/Al interfaces. (Miyamae et al. 2011) In OLEDs, Alq$_3$ is most widely used as electron transport/light emitting material. It is well known that Alq$_3$ has two possible geometrical isomers of meridional (C_1 symmetry) and facial (C_3 symmetry) forms, as shown in Fig. 1. In the meridional isomer, the three quinolate ligands around the central Al atom are not equivalent, while they are equivalent in the facial isomer. It has been reported that α- and amorphous Alq$_3$ consist of the meridional isomer, while γ- and δ-Alq$_3$ consist of facial isomer. (Nishiyama et al., 2009) Elucidating the electronic structures of the Alq$_3$/metal interface is required for its applicability to OLED. Such necessity should increase, due to the recent report of a significant enhancement of the current injection and OLED output induced by the insertion of an insulating layer such as LiF,(Hung et al., 1997) MgO, or MgF$_2$ (Shaheen et al., 1998) between the Al cathode and the Alq$_3$. Various mechanisms for this enhancement in the device efficiency have been proposed, and investigated using various techniques such as XPS, UPS, (Mason et al., 2001) and high resolution electron energy loss spectroscopy (HREELS). (He et al., 2000) One hypothesis is that thin LiF layer protects the Alq$_3$ from the deleterious reaction with Al.

Another hypothesis is that Li atoms produced by the dissociation of LiF by Al deposition lead to formation of the Alq_3 radical anion, is also considered. (Mason et al., 2001; Kido & Matsumoto 1998)

Figure 8 (a) shows the SFG spectra of 10 and 20 nm thick Alq_3 films on CaF_2 substrate excited by the visible light of 450 nm with the PPP polarization combination. In Fig. 8 (b), we compare the transmission IR spectrum of 69nm thick Alq_3 film deposited on CaF_2 substrate and the simulated IR spectrum for the Alq_3. The observed IR spectra of pristine Alq_3 agreed well with the IR spectrum of the meridional isomers in literature. (Kushto et al., 2000; Sakurai et al., 2004) The observed SFG spectra reasonably correspond to the IR spectrum in terms of the energies of the spectral features. Two peaks around 1600 cm^{-1} are derived from the C=C stretching modes of the quinolate ligands. The observed total SFG intensity of 20nm thick Alq_3 film shown in Fig. 8 was significantly larger than that of 10nm thick film, and the relative intensity of each peak also showed variation. We measure the SSP and PPP SFG spectra from 2 nm thick Alq_3 film deposited on CaF_2, however, the SFG signals derived from the Alq_3 were not observed.

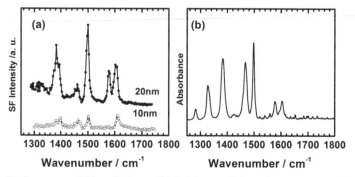

Fig. 8. (a) The SFG spectra of 10 and 20 nm thick Alq_3 on CaF_2 substrate excited by the visible light of 450 nm. (b) IR spectrum of 69 nm thick Alq_3 deposited on CaF_2 substrate.

In general, the growth of the SFG signal intensities with increasing the film thickness can be interpreted by the several reasons. First, the bulk SFG signal effect may come from the dipole-forbidden transitions. It is known that the dipole-forbidden transition observed in the SHG of the solid C_{60} films. (Koopmans et al., 1993; Kuhnke et al., 1998) Second-order nonlinearities of C_{60} are forbidden by symmetry under the electric-dipole approximation, but nonlocal bulk electric-quadrupole and magnetic-dipole processes appear due to its spherical character. However, such bulk contribution is not observed in the thickness dependence of the SHG on the photo-irradiated Alq_3 films. (Yoshizaki et al., 2005) Thus the bulk dipole-forbidden effects are excluded. Second, due to the existence of the two (air/Alq_3 and Alq_3/CaF_2) interfaces, the interference of the SFG signal between the two interfaces will induces the enhancements of the SFG signal intensities. The interference effects of the SFG signal due to the multiple layers are well analyzed by Hirose et al. (Hirose et al., 1998; Ishida et al., 1998) However, the SHG on the photo-irradiated Alq_3 films does not show such interference pattern. (Yoshizaki et al., 2005) Moreover, the decrease of in the intensity of the SHG signal was observed when the Alq_3 film was heated at a temperature lower than 100 °C. (Kajimoto et al., 2006) Although the part of the SFG signal might be influenced by the interference between the two interfaces, these SHG results suggest that the interference effect is not the main origin of the increase of the signal intensities. We attributed that the growth of the SFG signal is due to the non-centrosymmetric orientation of the bulk film, as in the case of the ordered formic acid layers on the Pt. (Hirose et al., 1998) It has been reported that a large potential is built in as-deposited Alq_3 thick film in dark and it decays rapidly by exposure to the ambient light observed by the Kelvin probe method. (Yoshizaki et al., 2005; Ito et al., 2002) According to the Kelvin probe experiments, (Ito et al., 2002) the surface potential increased rapidly, in the initial stage of Alq_3 deposition up to 1 nm on Al substrate, corresponding to the formation of the interfacial dipole layer. (Seki et al., 1997) After the initial shift, the surface potential increased linearly over a wide range of thickness from 5 to 550 nm. And this behaviour is independent on the kind of the substrate. (Hayashi et al., 2004) For Alq_3, meridional isomer possesses permanent dipole moment. Thus the permanent dipole moment of Alq_3 makes a significant contribution to the surface potential when these dipoles align unidirectionally in the film. On the other hand, the large surface potential is rapidly reduced by exposure to ambient light. (Yoshizaki et al., 2005; Ito et al., 2002) At first, this was considered with the photo-induced randomization of molecular orientation, (Sugi et al., 2004) however, the first-order electroabsorption measurements for the Alq_3 confirms that the non-centrosymmetric molecular orientation remains even after the light irradiation, indicating that the reduction of the large surface potential is not caused by the orientation randomization. (Isoshima et al., 2009) It is also suggested that the orientation polarization of Alq_3 film is maintained in OLED structure even under light illumination after the device fabrication. (Noguchi et al., 2008)

To gain further information about the thickness dependent SFG signal intensities, in Fig. 9 we show the SFG spectra of the Alq_3 films deposited on Au substrate excited by the visible light of 450 nm with the PPP polarization combination. In this experiments, we used the Au precoated Si substrates instead of the CaF_2 in order to investigate how thickness the non-centrosymmetric orientation in the bulk films appears from. As shown in Fig. 9, the significant increase of the SFG signal intensities are observed from the range of the thickness from 5 to 20 nm. This observation clearly indicates that the thickness of more than 5 nm thick Alq_3 film has uniaxial dipole orientation, because the SFG process is allowed in non-centrosymmetric media. In these spectra, it should be noted that the SFG peak shapes also shows the thickness dependence. This behavior must be mainly comes from the decrease in the intensity of the non-resonant contributions from the Au. It is known that the intensity of the non-resonant term of the SFG is changed by the coverage of the surface. (Himmelhaus et al., 2000) In contrast, the SFG spectra of the 1, 2 and 5 nm thick Alq_3 films deposited on Au does not show the significant thickness dependence, and these spectra show almost the same shapes and the same intensities. Since the Alq_3 thickness of 1 nm is comparable to the average thickness of the Alq_3 monolayers, (Yokoyama et al., 2003) these results indicates that the observed SFG peaks mainly come from the Alq_3/Au interface, and the molecular orientation of the inside of the Alq_3 films from the range of the thickness from 1 to 5 nm does not show the preferred orientation in the bulk. This finding is consistent with the previously reported thickness and the substrate dependence of the surface potential measurements for the Alq_3 evaporated films. (Hayashi et al., 2004) This result is also suggestive that the SFG signals from air/Alq_3 interface is negligibly small than those from the Alq_3/Au interface. SFG signal contribution from the air/Alq_3 interface will be further discussed in the next section. Consequently, we conclude that the inside of the thin films of Alq_3 does not form the effective non-centrosymmetric molecular orientation except the Alq_3/Au interface in the case of the film thickness under 5 nm.

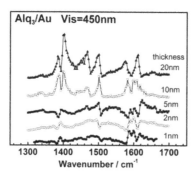

Fig. 9. Thickness dependence of the SFG spectra of Alq$_3$/Au.

As mentioned above, non-centrosymmetric orientation in the bulk films are remained even after the light irradiation, (Isoshima et al., 2009; Noguchi et al., 2008) while the surface potential observed in the Kelvin probe is rapidly reduced by the light-illumination. (Ito et al., 2002) The remaining of the molecular orientation on the bulk is consistent with our results of the thickness dependence of the SFG signal intensity. If the reduction of the surface potential is caused by the randomization of molecular orientation by the light-illumination, our probing lights should induce the randomization leading to much reduced SFG intensity and lack of thickness dependence in contrast to the observation, because the wavelength of the incident visible lights are near the absorption edge of Alq$_3$. Because the relation between the permanent dipole of the meridional Alq$_3$ and the vibrational transition dipole is not known, the mechanism for the appearance of the large surface potential in Alq$_3$ thick film is not elucidated by our SFG measurements. However, we conclude that the Alq$_3$ molecules spontaneously orient and form non-centrosymmetric orientation in the bulk even after the light irradiation in the case of the film thicknesses more than 5 nm.

4.2.2 DR-SFG study of Alq$_3$ thin films deposited on Al

In Fig. 10 we show the SFG vibrational spectra of the 2 nm thick Alq$_3$ deposited on Al with various visible wavelength in a PPP polarization (P-, P-, and P-polarized for SFG, visible, and IR, respectively) combination measured in the ambient condition. To minimize the influence of the air, the Alq$_3$ was covered with the 50 nm thick CaF$_2$. The SFG spectra of the 1 nm thick Alq$_3$/Al show almost the same to that of the 2 nm thick one. Since the film thickness of 1 nm is comparable to the average thickness of the Alq$_3$ monolayers, the observed SFG spectra of 2 nm thick Alq$_3$/Al are mainly originated from the Alq$_3$/Al interface. The 1 nm thick data is unstable and much worse reproducibility. On the other hand, the SFG spectra of the 5 nm thick Alq$_3$/Al systems show relatively stronger signals than those of the 1 and 2 nm thick Alq$_3$/Al. Such thickness dependent signal enhancements must be due to the effect of the uniaxial orientation of the molecular dipole, as mentioned in the preceding section. In order to minimize the effect of the uniaxial orientation, we used the 2 nm thick Alq$_3$ films for the SFG measurements. In the SFG spectra of the CaF$_2$/Alq$_3$/Al shows the bands at 1344, 1386, 1426, 1465, 1504, 1583, and 1612 cm^{-1}. The bands at 1583 and 1612 cm^{-1} are derived from the C=C stretching modes of the quinolate ligands. (Kushto et al., 2000; Sakurai et al., 2004) As shown in Fig. 10, remarkable changes in intensity of these peaks can be clearly observed by changing the visible wavelength. Figure 10 (b) shows the changes in the two representative peak strengths (A_l) of the peaks deduced from the fitting

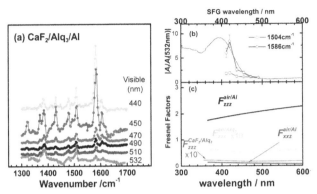

Fig. 10. (a) The SFG spectra of the Alq$_3$ on Al with various visible wavelengths. Spectra are offset for clarity. (b) Changes in the peak strengths of the 1583 and 1504 cm^{-1} as a function of the photon energies of the SFG. Optical absorption spectrum of the Alq$_3$ film is also shown by gray line. (c) Evolution of the Fresnel factors of F_{zzz} (black line) and F_{xxz} (red line) for the air/Al, F_{zzz} for the air/Alq$_3$ (green line), and F_{zzz} for the CaF$_2$/Alq$_3$ (blue line) interfaces. The F_{zzz}^{air/Alq_3} and $F_{zzz}^{CaF_2/Alq_3}$ are multiplied by 10.

of the DR-SFG spectra in Fig. 10 (a) using eq. (4) as a function of the photon energies of the SFG. In order to observe the enhancement ratios of the peak strength with the output SFG frequencies, the peak strengths are normalized with the strength of the SFG spectrum taken at the visible wavelength of 532 nm. For comparison, we show the optical absorption spectrum of the 20 nm thick Alq$_3$ film on CaF$_2$ in Fig. 10 (b). The excitation spectra exhibit a resonance almost coincident with the absorption spectrum for the Alq$_3$ thin film. Especially, significant increase in intensity is observed for the band at 1583 cm^{-1}. In contrast, the enhancements ratio of the SFG peak strength of the band at 1504 cm^{-1} is relatively weaker than that of the bands at 1583 cm^{-1}. These vibrational mode around 1583 cm^{-1} is assigned to the C=C stretching of the quinolate group while the 1504cm^{-1} peak is mixed modes that contain the contributions from the C-C and C-H in-plane bending motions. (Kushto et al., 2000) They are expected to have different degrees of coupling with the electronic transition. For Alq$_3$, the electronic transition at 390 nm is dominated by the π-π* excitation of the quinolate ligands. (Halls & Schlegel 2001) Thus it is reasonable that the C=C stretching of the quinolate ligands are effectively enhanced due to the resonance with the π-π* transitions. The SFG electronic excitation spectra in Fig. 10 (b) are slightly shifted to lower frequency as compared to the optical absorption spectrum of the Alq$_3$. The shift of the electronic transition peak may be suggestive that the electronic excitation gap at the interface becomes smaller than that of the bulk. Since the SFG excitation spectra are not measured in the whole region across the optical transition peak, and therefore, further experiments with the shorter wavelength excitation are needed to reveal the definitive information of the excitation profiles at the interface.

For the analysis of the doubly-resonant SFG, it should be important to note that the changes of the Fresnel factors have to be considered since it might change with the variation of the visible wavelength. The effective the second-order nonlinear susceptibility tensor components of an azimuthally isotropic sample contributes to the PPP SFG signals can be written as

$$
\begin{aligned}
A_{q,PPP} = &-L_{xx}(\omega_{SF})L_{xx}(\omega_{VIS})L_{zz}(\omega_{IR})\cos\beta_{SF}\cos\beta_{VIS}\sin\beta_{IR}\chi_{xxz} \\
&-L_{xx}(\omega_{SF})L_{zz}(\omega_{VIS})L_{xx}(\omega_{IR})\cos\beta_{SF}\sin\beta_{VIS}\cos\beta_{IR}\chi_{xzx} \\
&+L_{zz}(\omega_{SF})L_{xx}(\omega_{VIS})L_{xx}(\omega_{IR})\sin\beta_{SF}\cos\beta_{VIS}\cos\beta_{IR}\chi_{zxx} \\
&+L_{zz}(\omega_{SF})L_{zz}(\omega_{VIS})L_{zz}(\omega_{IR})\sin\beta_{SF}\sin\beta_{VIS}\sin\beta_{IR}\chi_{zzz}
\end{aligned}
\qquad (7)
$$

where $L_{xx}(\omega)$ and $L_{zz}(\omega)$ are the Fresnel coefficients at frequency ω; β_{SF}, β_{VIS}, and β_{IR} are the reflection angles of the sum frequency, visible, and IR pulses, respectively; and χ_{ijk}s are the nonvanishing elements of the second-order nonlinear susceptibility. (Zhuang et al., 1999) We found that χ_{xzx} and χ_{zxx} are much smaller than χ_{xxz} and χ_{zxx}. Thus the Fresnel factors $F_{zzz} = \left| L_{ZZ}(\omega_{SF})L_{ZZ}(\omega_{VIS})L_{ZZ}(\omega_{IR})\sin\beta_{SF}\sin\beta_{VIS}\sin\beta_{IR}\right|$ and $F_{xxz} = \left| L_{XX}(\omega_{SF})L_{XX}(\omega_{VIS})L_{ZZ}(\omega_{IR})\cos\beta_{SF}\cos\beta_{VIS}\sin\beta_{IR}\right|$ were calculated using the complex refractive indices of metallic aluminum. (Parik, 1985) As shown in Fig. 10 (b), the F_{zzz} at air/Al interface monotonically decreases with the increase of the photon energy of the SFG, and it does not explain the evolution of the SFG intensities of the Alq$_3$/Al. Thus we conclude that the changes of the Fresnel factors do not much affect on the spectral shape of the SFG excitation profile of the Alq$_3$/Al.

The analysis of the Fresnel factor is also important for the investigation of such a thin layered sample. We calculated the Fresnel factors F_{zzz} at air/Alq$_3$ and CaF$_2$/Alq$_3$ interfaces, assuming that the air/Alq$_3$ interface is azimuthally isotropic. The wavelength dependence of the refractive indices for Alq$_3$ were used for the evaluation of the Fresnel factors at air/Alq$_3$ and CaF$_2$/Alq$_3$ interfaces. (Djurišić et al., 2002) As shown in Fig. 10 the F_{zzz} at air/Alq$_3$ and CaF$_2$/Alq$_3$ interfaces are much smaller than the F_{zzz} at air/Al interface. From this observation, we conclude that the experimentally observed SFG should be mainly from the Alq$_3$/Al interface.

In the previous UPS study of the Alq$_3$ deposited on Al systems, an extra occupied state above the HOMO level was detected, suggesting a strong chemical interaction between Alq$_3$ and Al. (Yokoyama et al., 2003) It was suggested that the interaction between Al and Alq$_3$ is somewhat different from the charge transfer as reported for alkaline metal doped Alq$_3$. (Curioni & Andreoni, 1999) On the other hand, a theoretical calculation of Alq$_3$ layer on Al (111) suggested that the interfacial interaction is weak, (Yanagisawa & Morikawa 2006) These previous findings are controversial and not easy to discuss consistently, but one possible reason of this discrepancy must be due to the least reactivity of the clean Al(111) surface used in the DFT. (Yanagisawa & Morikawa 2008) If the strong chemical interaction occurs at the interface, the vibrational SFG spectra and the corresponding electronic excitation profiles should provide different behavior to those of the bulk. On the contrary, the SFG excitation profile of the 1583 cm^{-1} band is almost identical to the visible optical absorption spectrum of the pristine Alq$_3$, and we cannot find such spectral changes. Although the electronic excitation profile does not show significant changes, however, the position of the C=C stretching peak shows slightly red-shifted. As discussed below, red-shift of the C=C peak is indicative of the charge transfer from the Al substrate to Alq$_3$.

4.2.3 DR-SFG study of Al layer deposited onto Alq$_3$

The traditional surface analysis techniques such as UPS have been applied to examine the interfaces formed by depositing organic material on metals, which are not much troubled by the factor of chemical reaction. (Ishii et al., 1999) Actually, in many cases the UPS spectra of organic-on-metal systems show only rigid shifts on the energy scale, suggesting the absence of strong chemical interaction. In contrast, actual OLEDs are fabricated by organic layers sandwiched by a cathode and anode. The buried interface between metal anode and the organic layer is formed by the deposition of the metal on organic materials. When the metal is deposited on an organic layer by evaporation, the high reactivity of the vaporized hot metal atom often leads to a chemical reaction at the interface, and diffuse into the organic layer. Due to the energy transfer from the hot metal atom, deposition of the metal on organic materials may also induce the decomposition of the molecules, polymerization of the molecules, reorientation of the molecules, desorption of the organic materials from the substrate, and so on. Thus the "*metal-on-organic*" systems are generally much more complex than the organic-on-metal systems. Due to the high reactivity of the Al atom, reactions between the Alq_3 and Al are expected by the deposition of the Al layer. For the characterization of the buried metal-on-organic interface, next, we measure the SFG spectra for the Al deposited on Alq_3 film. The deposited thick Al layer can act as superior gas barrier, and the extent of the oxidation of the Al interface is much reduced.

In Fig. 11, we show the SFG spectra of Al film directly deposited on 5nm thick Alq_3 film on CaF_2 substrate with various visible wavelengths in a PPP polarization. In the case of the $Al/Alq_3/CaF_2$ system, the thickness of Alq_3 is set to 5 nm. When we use the Alq_3 layer of 2 nm thick sandwiched by CaF_2 and Al, the SFG gives quite weak signal. Low sticking probability of the Alq_3 on CaF_2 is not plausible judging from the molecular weight of Alq_3. Although the exact reason is not clear at this present, we tentatively thought that the detectable interface could not be formed by the deposition of the 2 nm thick Alq_3. As mentioned above, the metal atoms often diffuse into the organic layer, and this process may prevent to form the clear interface for the 2 nm thick Alq_3 sandwiched between CaF_2 and Al. It should be noted that the SFG signal comes mainly from the Al/Alq_3 interface, not from the Alq_3/CaF_2 interface, because no SFG signals are detected from the thin Alq_3 layer deposited on CaF_2. This is further supported by the Fresnel factor difference between Al and CaF_2 interfaces. We show in Fig. 10 (b), the F_{zzz} at CaF_2/Alq_3 interface are negligibly smaller

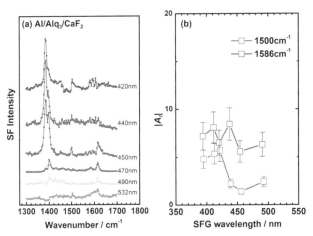

Fig. 11. (a) The SFG spectra of $Al/Alq_3/CaF_2$ substrate. (b) Changes in the peak strengths of the 1586 and 1500 cm^{-1} of the SFG spectra as a function of the photon energies of the SFG.

than F_{zzz} at air/Al interface. One may think that the SFG spectra in Fig. 11 are comes from the Alq$_3$ bulk, since the thick Alq$_3$ film shows uniaxial orientation. Although the evaluation of the bulk contribution needs transmission experiments, (Wei et al., 2000) if the SFG signals in Fig. 11 are mainly originated from the bulk Alq$_3$, the spectral shapes and their behavior should be similar to those of the Alq$_3$ on Al systems. Therefore, we thought that the main contribution of the SFG signals is the Al/Alq$_3$ interface.

The spectral features are different from the SFG spectra of Alq$_3$/Al and the pristine IR spectrum of the Alq$_3$. The relative intensities of bands derived from the C=C stretching around 1600 cm^{-1} becomes weak as compared to the case of the Alq$_3$/Al. Figure 11 (b) shows the changes in the two representative peak strengths (A_l) of the vibrational peaks deduced from the fitting of the DR-SFG spectra in Fig. 11 (a) as a function of the photon energies of the SFG. As shown in Fig. 11 (b), the SFG electronic profiles derived from the C=C bands does not agree with the linear optical absorption spectrum of the Alq$_3$, indicating that the electronic-resonant effects associated with the π-π^* transitions in the quinolate rings are almost vanished. According to the theoretical calculations for the Al-Alq$_3$ complex, the energy diagrams near the gaps are significantly changed by the chemical bonding formation between Al and Alq$_3$. (Curioni & Andreoni, 1999) In contrast to the case of the alkaline-metal-Alq$_3$ complex, HOMO of pristine Alq$_3$ is destabilized by the interaction with the Al 3s orbital. The interaction with the Al is such that one of the Alq$_3$ HOMOs is repelled to higher energy, and a state with predominant Al character appears in the same energy range. Al-Alq$_3$ interaction also induces the modification of LUMO. Previous NEXAFS study of Al/Alq$_3$ interface suggests that the Al-Alq$_3$ interaction is not simple electron transfer from Al to Alq$_3$. (Yokoyama et al., 2005) Although the theoretical simulations for the 1:1 Al-Alq$_3$ complex cannot predict the observed NEXAFS results, modification of HOMO (LUMO) level occurs at the Al/Alq$_3$ interface. Thus we conclude that disappearance of the doubly-resonant effect associated with the π-π^* transitions must be caused by the perturbation of the HOMO and LUMO of Alq$_3$ by the interaction of the Al.

4.2.4 Effects of the LiF insertion between Al and Alq$_3$

In this section, we discuss the effects of the insertion of a LiF layer between Al and Alq$_3$ interface. The SFG spectra of the Al/LiF/Alq$_3$ system with various visible wavelengths in a PPP polarization are shown in Fig. 12 (a). The spectral features are quite different from those of Alq$_3$/Al and Al/Alq$_3$. The new broad bands, which show the weak excitation wavelength dependence, appear around 1335 and 1450 cm^{-1}. The C=C stretching modes of the quinolate ligands are observed at 1572 and 1607 cm^{-1}. The frequency shift to lower wavenumber of the C=C stretching mode is also reported in the IR and DFT study of the potassium-doped Alq$_3$. (Sakurai et al., 2004) The DFT calculations and the IR spectrum for the potassium-doped Alq$_3$ suggested that the C=C stretching frequency of Alq$_3$ anion is lower than that of pristine Alq$_3$ molecule. Consequently, the red-shift of the C=C stretching bands in the SFG spectra is indicative of the formation of the Alq$_3$ anionic states upon reaction with Li at the interface. (Mason et al., 2001) This observation is in good agreement with the previous UPS and XPS measurements for the Al/LiF/Alq$_3$ interfaces (Mason et al., 2001; Ding et al., 2009) and, to our knowledge, this is the first observation of the anionic state formation at the buried interface in ambient condition. Figure 12 (b) shows the SFG excitation profiles obtained from the DR-SFG spectra in Fig. 12 (a) as a function of the photon energies of the SFG. The SFG excitation profiles derived from the 1572 cm^{-1} band gives maximum around 420 nm, however, it does not show large shift to the lower photon energy. If the Alq$_3$ at the LiF interface forms the Alq$_3$ anion, the absorption peaks should appear below 600 nm. (Ganzorig & Fujihira, 2002) The charges transferred from the Li might smear the SFG excitation profiles of the C=C stretching. Unfortunately, our SFG system cannot generate the sufficient power of the light below 640 nm at present. Further experiments with the longer wavelength excitation will reveal the electronic character at the charged interface.

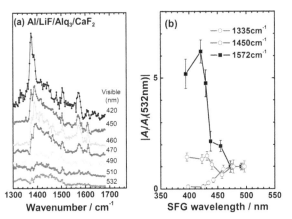

Fig. 12. (a) The SFG spectra of Al/LiF/Alq$_3$. (b) The changes in the peak strengths of the three representative vibrational peaks as a function of the photon energies of the SFG.

Next, we mention the origin of the broad peaks at 1335 and 1450 cm^{-1}. Similar broad features are observed by IR and Raman studies of potassium-doped Alq$_3$ and the Al/LiF/Alq$_3$ system. (Sakurai et al., 2004) It was suggested that these bands are derived from the quinolate ligands reacted with potassium or lithium. On the other hand, recent Raman studies performed in UHV environment show the similar broad features around 1355, 1405, and 1560 cm^{-1} for Alq$_3$/Al system, and 1315, 1435, and 1515 cm^{-1} for Alq$_3$/Ca system, respectively. (Davis & Pemberton, 2008, 2009) They assigned these modes are derived from the G-bands and the D-bands of the graphitic carbon generated by the deposition of Ca or Al onto Alq$_3$ thick film. One may think that the graphite related bands are IR inactive modes, however, these modes become IR active in the nitrogenated amorphous graphite. (Kaufman et al., 1989) NEXAFS studies also suggest the existence of the chemical interaction different from electron transfer at the Al/LiF/Alq$_3$ system, which differs from Al/Alq$_3$ and Li/Alq$_3$ interfaces. (Yokoyama et al., 2005) Although further studies are necessary to determine the origin of these broad features, the wavelength independent SFG excitation profiles of these features may be suggestive that these bands are originated from the graphite-like bands, not from the Alq$_3$. On the other hands, the red-shift of the other SFG peaks, which shows the remarkable wavelength dependent SFG excitation profiles, may be indicative by the formation of the Alq$_3$ anionic states at the interface, as mentioned above. Because the interface formed by the metal deposition onto the organic materials is much more complicated than the organic-on-metal system, we conclude that the Al/LiF/Alq$_3$ buried interface might be co-existence of the negatively charged Alq$_3$ by the charge transfer from the Li and the Li-reacted graphitic carbon-like Alq$_3$.

Finally, we discuss the interfacial vibrational and electronic structural difference between Al/Alq3 and Al/LiF/Alq3 interfaces by comparing with the SFG spectra. Figure 13 shows the SFG spectra of the Alq3/Al, Al/Alq3, and Al/LiF/Alq3 systems excited by the visible light of 450 nm with the PPP polarization combination. From these data, we can confirm three differences by the different interface. First, the red-shift of the C=C stretching modes are observed in the Al/LiF/Alq3 interfaces, while such shift is not observed in the case of the Al/Alq3 system. These spectral behaviors clearly indicate that the chemical interaction between Li and Alq3 occurred at the Al/LiF/Alq3 interface. Red-shift of the SFG vibrational modes is suggestive to the formation of the Alq3 anionic states at the interface. Second, the electronic-resonance for the C=C stretching modes are vanished for the Al/Alq3 system, while it remains in the case of the Al/LiF/Alq3 system. This observation indicates that the energy diagrams near the band gaps are modified by the chemical reaction between Al and Alq3, rather than the simple charge transfer from Al to Alq3. Finally, the broad features which might be due to the graphite-like carbon are observed in the case of the Al/LiF/Alq3 system. Insertion of the LiF layer between Al and Alq3 induces reaction at the interface, and it emerges the negatively charged Alq3 and the Li-reacted graphitic carbon-like Alq3.

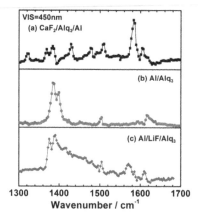

Fig. 13. The SFG spectra of the (a) CaF2/Alq3/Al, (b) Al/Alq3, and (c) Al/LiF/Alq3 systems excited by the visible light of 450 nm with the PPP polarization combination.

5. Conclusion

The interfacial vibrational and electronic states of organic/metal interfaces were studied using doubly-resonant SFG spectroscopy. In the SFG studies of the air/PFO and the Al/LiF/PFO interfaces, the planes of PFO rings are nearly parallel to the surface plane at the air/PFO interface. This planar orientation induced by the interface confinement effects leads to the smaller band gaps at the air/PFO and the Al/LiF/PFO interfaces than those of the bulk PFO. In the DR-SFG spectra of the Alq3/Al, remarkable changes in intensities of the SFG peaks derived from the C=C stretching of the quinolate ligands can be clearly observed by changing the visible wavelength due to the double resonance effect. In contrast, The SFG excitation profiles of the C=C stretching of the Al/Alq3 interfaces do not show the wavelength dependences, indicating that the perturbation of the HOMO and LUMO of pristine Alq3 by the interaction of the Al. The SFG spectra of the Al/LiF/Alq3 system indicates that the existence of the Alq$_3$ anionic states at the buried interface in atmospheric condition. Additional broad bands around 1335 and 1450 cm^{-1} might be due to the existence of the Li-reacted graphitic carbon-like Alq$_3$. Co-existence of the negatively charged Alq$_3$ and the Li-reacted graphitic carbon-like Alq$_3$ at the buried interface is proposed.

Although the experimental results obtained here were performed in ambient condition, the present study lead to a better understanding of the molecular and the electronic structure of the organic/metal interfaces. In much the same way as multi-dimensional spectroscopy is suitable for studies of intra- and inter-molecular interactions in the bulk, the DR-SFG provides a similar opportunity for studies of molecules at interfaces. This doubly-resonant SFG technique, which was proved feasible for OLEDs study in this chapter, offers a novel spectroscopy for the characterization of the vibrational and electronic structures at the buried organic interfaces.

6. Acknowledgments

The work in this paper was performed in collaboration with Professors H. Ishii, Y. Ouchi, and K. Seki, and Drs W. Mizutani, K. Tsukagoshi, Y. Noguchi, E. Ito, and Y. Sakurai; all are gratefully acknowledged. We gratefully acknowledge financial support from the Mitsui Chemicals Inc. This work was supported in part by Grants-in-Aid for Scientific Research from the Japan Society for the Promotion of Science (JSPS) of Japan.

7. References

Ariu, M.; Lidzey, D. G.; Bradley, D. D. C., Influence of film morphology on the vibrational spectra of dioctyl substituted polyfluorene (PFO), Synth. Met., Vol. 111 (2000), pp. 607-610.

Becker, K.; Lupton, J. M., Dual species emission from single polyfluorene molecules: signatures of stress-induced planarization of single polymer chains, J. Am. Chem. Soc., Vol. 127, No. 20 (2005), pp. 7306-7307.

Burroughes, J. H.; Bradley, D. D. C.; Brown, A. R.; Marks, R. N.; Mackay, K.; Friend, R. H.; Burn, P. L.; Holmes, A. B., Light-emitting diodes based on conjugated polymers, Nature, Vol. 347, No. 6293 (1990), pp. 539-540.

Cadby, A. J.; Lane, P. A.; Mellor, H.; Martin, S. J.; Grell, M.; Giebeler, C.; Bradley, D. D. C.; Wohlgenannt, M.; An, C.; Vardeny, Z. V., Film morphology and photophysics of polyfluorene, Phys. Rev. B, Vol. 62, No. 23 (2000), 62, pp. 15604-15609.

Campoy-Quiles, M.; Heliotis, G.; Xia, R.; Ariu, M.; Pintani, M.; Etchegoin, P.; D. D. C. Bradley, D. D. C., Ellipsometric characterization of the optical constants of polyfluorene gain media, Adv. Func. Mater., Vo. 15, No. 6 (2005), pp. 925-933.

Cheum, H.; Galbrecht, F.; Nehls, B.; Scherf, U.; Winokur, M. J., Interface specific variations in the steady-state optical properties of polyfluorene thin films, J. Mater. Sci.: Mater. Electronics, Vol. 20, Supplement 1 (2009), S498-S504.

Curioni, A.; Andreoni, W., Metal−Alq$_3$ complexes: The nature of the chemical bonding, J. Am. Chem. Soc., Vol. 121, No. 36 (1999), pp. 8216-8220.

Davis, R. J.; Pemberton, J. E., Investigation of the interfaces of tris-(8-hydroxyquinoline) aluminum with Ag and Al using surface Raman spectroscopy, *J. Phys. Chem. C*, Vo. 112, No. 11 (2008), pp. 4364-4371.

Davis, R. J.; Pemberton, J. E., Surface Raman spectroscopy of chemistry at the tris(8-hydroxyquinoline) aluminum/Ca interface, *J. Phys. Chem. A*, Vol. 113, No. 16 (2009), pp. 4397-4402.

Ding, H.; Park, K.; Gao, Y.; Kim, D. Y.; So, F., Electronic structure and interactions of LiF doped tris-(8-hydroxyquinoline) aluminum (Alq), *Chem. Phys. Lett.*, Vol. 473, Nos. 1-3 (2009), pp. 92-95.

Djurišič, A. B.; Kwong, C. Y.; Guo, W. L.; Lau, T. W.; Li, E. H.; Liu, Z. T.; Kwok, H. S.; Lam, L. S. M.; Chan, W. K., Spectroscopic ellipsometry of the optical functions of *tris*-(8-hydroxyquinoline) aluminum (Alq3),*Thin Solid Films*, Vol. 416, Nos. 1-2 (2002), pp. 233-241.

Ganzorig, C.; Fujihira, M., A possible mechanism for enhanced electrofluorescence emission through triplet–triplet annihilation in organic electroluminescent devices, *Appl. Phys. Lett.*, Vol. 81, No 17 (2002), pp. 3137-3179.

Gong, X.; Iyer, P. K.; Moses, D.; Bazon, G. C.; Heeger, A. J.; Xiao, S. S., Stabilized blue emission from polyfluorene-based light-emitting diodes: Elimination of fluorenone defects, *Adv. Func. Mater.*, Vol. 13, No. 4 (2003), pp. 325.

Grell, M.; Bradley, D. D. C.; Ungar, G.; Hill, J.; Whitehead, K. S., Interplay of physical structure and photophysics for a liquid crystalline polyfluorene, *Macromolecules*, Vol. 32, No. 18 (1999), pp. 5810-5817.

Halls, M. D.; Schlegel, H. B., Molecular orbital study of the first excited state of the OLED material tris-(8-hydroxyquinoline)aluminum(III), *Chem. Mater.*, Vol. 13, No. 8 (2001), pp. 2632-2640.

Hayashi, M.; Lin, S. H.; Raschke, M. B.; Shen, Y. R., A Molecular theory for doubly resonant IR–UV-vis sum-frequency generation, *J. Phys. Chem. A*, Vol. 106, No. 10 (2002), pp. 2271-2282.

Hayashi, N.; Imai, K; Suzuki, T.; Kanai, K.; Ouchi, Y.; Seki, K., Substrate dependence of giant surface potential of Alq3 and the examination of surface potential of related materials, *Proceedings of International Symposium on Super-Functionality Organic Devices*, IPAP Conf. Series 6, pp. 69-72, ISBN4-900526-20-7 Chiba, October, 2004.

He, P.; Au, F. C. K.; Wang, Y. M.; Cheng, L. F.; Lee, C. S.; Lee, S. T., Direct evidence for interaction of magnesium with tris(8-hydroxy-quinoline) aluminum, *Appl. Phys. Lett.*, Vol. 76, No. 11 (2000), pp. 1422-1424.

Himmelhaus, M.; Eisert, F.; Buck, M.; Grunze, M., Self-assembly of *n*-alkanethiol monolayers; A study by IR–visible sum frequency spectroscopy (SFG), *J. Phys. Chem. B*, Vol. 104, No. 3 (2000), pp. 576-584.

Hirose, C.; Ishida, H.; Iwatsu, K.; Watanabe, N.; Kubota, J.; Wada, A.; Domen, K., In situ SFG spectroscopy of film growth. I. General formulation and the analysis of the signal observed during the deposition of formic acid on Pt(110)-(1×2) surface, *J. Chem. Phys.*, Vol. 108, No. 14 (1998), pp. 5948-5956.

Hung, L. S.; Tang, C. W.; Mason, M. G., Enhanced electron injection in organic electroluminescence devices using an Al/LiF electrode, *Appl. Phys. Lett.*, Vol. 70, No. 2 (1997), pp. 152-154.

Ishida, H.; Iwatsu, K.; Kubota, J.; Wada, A.; Domen, K.; Hirose, C., *In situ* SFG spectroscopy of film growth. II. Deposition of formic acid on Ni(110) surface, *J. Chem. Phys.*, Vol. 108, No. 14 (1998), pp. 5957-5964.

Ishii, H.; Sugiyama, K.; Ito, E.; Seki, K., Energy level alignment and interfacial electronic structures at organic/metal and organic/organic interfaces, *Adv. Mater.*, Vol. 11, No. 8 (1999), pp. 605-625.

Isoshima, T.; Ito, H.; Ito, E.; Okabayashi, Y.; Hara, M., Long-term relaxation of molecular orientation in vacuum-deposited Alq$_3$ thin films, *Mol. Cryst. Liq. Cryst.*, Vol. 505, No. 1 (2009), pp. 59-63.

Ito, E.; Washizu, Y.; Hayashi, N.; Ishii, H.; Matsuie, N.; Tsuboi, K.; Harima, Y.; Yamashita, K.; Seki, K., Spontaneous buildup of giant surface potential by vacuum deposition of Alq$_3$ and its removal by visible light irradiation, *J. Appl. Phys.*, Vol. 92, No. 12 (2002), pp. 7306-7310.

Iwahashi, T.; Miyamae, T.; Kanai, K.; Seki, K.; Kim, D.; Ouchi, Y., Anion configuration at the air/liquid interface of ionic liquid [bmim]OTf studied by sum-frequency generation spectroscopy, *J. Phys. Chem. B*, Vol. 112, No. 38 (2008), pp. 11936-11941.

Kajimoto, N.; Manaka, T.; Iwamoto, M., Decay process of a large surface potential of Alq$_3$ films by heating, *J. Appl. Phys.*, Vol. 100, No. 5 (2006), pp. 053707.

Kaufman, J. H.; Metin, S.; Saperstein, D. D., Symmetry breaking in nitrogen-doped amorphous carbon: Infrared observation of the Raman-active G and D bands, *Phys. Rev. B*, Vol. 39, No. 18 (1989), pp. 13053-13060.

Kawana, S.; Durrel, M.; Lu, J.; Macdonald, J. E.; Grell, M.; Bradley, D. D. C.; Jukes, P. C.; Jones, R. A. L.; Bennett, S. L., X-ray diffraction study of the structure of thin polyfluorene films, *Polymer*, Vol. 43, No. 6 (2002), pp. 1907-1913.

Kido, J.; Matsumoto, T., Bright organic electroluminescent devices having a metal-doped electron-injecting layer, *Appl. Phys. Lett.*, Vol. 73, No. 20 (1998), pp. 2866-2868.

Knaapila, M.; Lyons, B. P.; Kisko, K.; Foreman, J. P.; Vainio, U.; Mihaylova, M.; Seeck, O. H.; Palsson, L. O.; Serimaa, R.; Torkkeli, M.; Monkma, A. P., X-ray diffraction studies of multiple orientation in poly(9,9-bis(2-ethylhexyl)fluorene-2,7-diyl) thin films, *J. Phys. Chem. B*, Vol. 107, No. 45 (2003), pp. 12425-12430.

Koopmans, B.; Janner, A.-M.; Jonkman, H. T.; Sawatzky, G. A.; van der Woude, F., Strong bulk magnetic dipole induced second-harmonic generation from C$_{60}$, *Phys. Rev. Lett.*, Vol. 71, No. 21 (1993), pp. 3569-3572.

Kuhnke, K.; Epple, M.; Kern, K., Second-harmonic spectroscopy of fullerenes, *Chem. Phys. Lett.*, Vol. 294, Nos. 1-3 (1998), pp. 241-247.

Kushto, G. P.; Iizumi, Y.; Kido, J.; Kafafi, Z. H., A matrix-isolation spectroscopic and theoretical investigation of tris(8-hydroxyquinolinato)aluminum(III) and tris(4-methyl-8-hydroxyquinolinato)aluminum(III), *J. Phys. Chem. A*, Vol. 104, No. 16, (2000), pp. 3670-3680.

Li, Q.; Hua, R.; Chou, K. C., Electronic and conformational properties of the conjugated polymer MEH-PPV at a buried film/solid interface investigated by two-dimensional IR−visible sum frequency generation, *J. Phys. Chem. B*, Vol. 112, No. 8 (2008), pp. 2315-2318.

List, E. J. W.; Guentner, R., The effect of keto defect sites on the emission properties of polyfluorene-type materials, *Adv. Mater.*, Vol. 14, No. 5, (2002), pp. 374-378.

Mallavia, R.; Montilla, F.; Pastor, I.; Velasquez, P.; Arredondo, B.; Alvarez, A. L.; Mateo, C. R., Characterization and side chain manipulation in violet-blue poly-[(9,9-dialkylfluoren-2,7-diyl)-*alt*-co-(benzen-1,4-diyl)] backbones, *Macromolecules*, Vol. 38, No. 8 (2005), pp. 3185-3192.

Mason, M. G.; Tang, C. W.; Hung, L. -S.; Raychaudhuri, P.; Madathi, J.; Giesen, D. J.; Yan, L.; Le, Q. T.; Gao, Y.; Lee, S. -T.; Liao, L. S.; Cheng, L. F.; Salaneck, W. R.; dos Santos, D. A.; Brédas, J. L., Interfacial chemistry of Alq$_3$ and LiF with reactive metals, *J. Appl. Phys.*, Vol. 89, No. 5 (2001), pp. 2756-2765.

Miyamae, T.; Nozoye, H., Correlation of molecular conformation with adhesion at AlO$_x$/poly(ethylene terephthalate) interface studied by sum-frequency generation spectroscopy, *Appl. Phys. Lett.*, Vol. 85, No. 19 (2004), pp. 4373-4375 (2004).

Miyamae, T.; Akiyama, H.; Yoshida, M.; Tamaoki, N., Characterization of poly(*N*-isopropylacrylamide)-grafted interfaces with sum-frequency generation spectroscopy, *Macromolecules*, Vol. 40, No. 13 (2007), pp. 4601-4606.

Miyamae, T.; Morita, A.; Ouchi, Y., First acid dissociation at an aqueous H$_2$SO$_4$ interface with sum frequency generation spectroscopy, *Phys. Chem. Chem. Phys.*, Vol. 10, No. 15 (2008), pp. 2010-2013.

Miyamae, T.; Miyata, Y.; Kataura, H., Two-color sum-frequency generation study of single-walled carbon nanotubes on silver, *J. Phys. Chem. C*, Vol. 113, No. 34, (2009), pp. 15314-15319.

Miyamae, T.; Tsukagoshi, K.; Mizutani, W., Two-color sum frequency generation study of poly(9,9-dioctylfluorene)/electrode interfaces, *Phys. Chem. Chem. Phys.*, Vol. 12, No. 44 (2010), pp. 14666-14669.

Miyamae, T.; Ito, E.; Noguchi, Y.; Ishii, H., Characterization of the interactions between Alq$_3$ thin films and Al probed by two-color sum-frequency generation spectroscopy, *J. Phys. Chem. C*, Vol. 15, No. 19 (2011), pp. 9551-9560.

Montilla, F.; Mallavia, R., On the origin of green emission bands in fluorene-based conjugated polymers, *Adv. Func. Mater.*, Vol. 17, No. 1 (2007), pp. 71-78.

Nishiyama, Y.; Fukushima, T.; Takami, K.; Kusaka, Y.; Yamazaki, T.; Kaji, H., Characterization of local structures in amorphous and crystalline tris(8-hydroxyquinoline) aluminum(III) (Alq$_3$) by solid-state ^{27}Al MQMAS NMR spectroscopy, *Chem. Phys. Lett.*, Vol. 471, Nos. 1-3 (2009), pp. 80-84.

Noguchi, Y.; Sato, N.; Tanaka, Y.; Nakayama, Y.; Ishii, H., Threshold voltage shift and formation of charge traps induced by light irradiation during the fabrication of organic light-emitting diodes, *Appl. Phys. Lett.*, Vol. 92, No. 20 (2008), pp. 203306.

Parik, E. D., Ed. *Handbook of Optical Constants of Solids*, 1985, Academic Press, ISBN 0-12-544420-6, London.

Raschke, M. B.; Hayashi, M.; Lin, S. H.; Shen, Y. R., Doubly-resonant sum-frequency generation spectroscopy for surface studies, *Chem. Phys. Lett.*, Vol. 359, Nos. 5-6 (2002) pp.367-372.

Sakurai, Y.; Hosoi, Y.; Ishii, H.; Ouchi, Y.; Salvan, G.; Kobitski, A.; Kampen, T. U.; Zahn, D. R. T.; Seki, K., Study of the interaction of tris-(8-hydroxyquinoline) aluminum (Alq$_3$) with potassium using vibrational spectroscopy: Examination of possible isomerization upon K doping, *J. Appl. Phys.*, Vol. 96, No. 10 (2004), pp. 5534-5542.

Salaneck, W. R.; Seki, K.; Kahn, A.; Pireaux, J. J., (Eds.) 2002, *Conjugated Polymer and Molecular Interfaces – Science and Technology for Photonic and Optoelectronics Applications*, Marcel Dekker, ISBN 0-8247-0588-2, New York.

Seki, K.; Ito, E.; Ishii, H., Energy level alignment at organic/metal interfaces studied by UV photoemission, *Synth. Met.*, Vol. 91, Nos. 1-3 (1997), pp. 137-142.

Shaheen, S. E.; Jabbour, G. E.; Morrell, M. M.; Kawabe, Y.; Kippelen, B.; Peyghambarian, N.; Nabor, M. -F.; Schlaf, R.; Mash, E. A.; Armstrong, N. R., Bright blue organic light-emitting diode with improved color purity using a LiF/Al cathode, *J. Appl. Phys.*, Vol. 84, No. 4 (1998), pp. 2324-2327.

Shen, Y. R., November 2002, *The principles of Nonlinear Optics*: Wiley, ISBN: 978-0-471-43080-3, New York.

Silva, H. S.; Miranda, P. B., Molecular ordering of layer-by-layer polyelectrolyte films studied by sum-frequency vibrational spectroscopy, *J. Phys. Chem. B*, Vol. 113, No. 30 (**2009**), pp. 10068–10071.

Sugi, K.; Ishii, H.; Kimura, Y.; Niwano, M.; Ito, E.; Washizu, Y.; Hayashi, N.; Ouchi, Y.; Seki, K., Characterization of light-erasable giant surface potential built up in evaporated Alq_3 thin films, *Thin Solid Films*, Vol. 464-465 (2004), pp. 412-415.

Tang, C. W.; Slyke, S. A. V, Organic electroluminescent diodes, *Appl. Phys. Lett.*, Vol. 51, No. 12 (1987), pp. 913-915.

Wei, X.; Hong, S. C.; Lvovsky, A. I.; Held, H.; Shen, Y. R., Evaluation of surface vs bulk contributions in sum-frequency vibrational spectroscopy using reflection and transmission geometries, *J. Phys. Chem. B*, Vol. 104, No. 14 (2000), pp. 3349-3354.

Wu, D.; Deng, G. H.; Guo, Y.; Wang, H. F., Observation of the interference between the intramolecular IR–visible and visible–IR processes in the doubly resonant sum frequency generation vibrational spectroscopy of rhodamine 6G adsorbed at the air/water interface, *J. Phys. Chem. A*, Vol. 113, No. 21 (2009), pp. 6058-6063.

Yanagisawa, S; Morikawa, Y., Theoretical investigation on the electronic structure of the tris-(8-hydroxyquinolinato) aluminum/aluminum interface, *Jpn. J. Appl. Phys.*, Vol. 45, No. 1B (2006), pp. 413-416.

Yanagisawa, S.; Lee, K.; Morikawa, Y., First-principles theoretical study of Alq_3/Al interfaces: Origin of the interfacial dipole, *J. Chem. Phys.*, Vol. 128, No. 24 (2008), pp. 244704.

Yokoyama, T.; Yoshimura, D.; Ito, E.; Ishii, H.; Ouchi, Y.; Seki, K., Energy level alignment at Alq_3/LiF/Al interfaces studied by electron spectroscopies: Island growth of LiF and size-dependence of the electronic structures, *Jpn. J. Appl. Phys.*, Vol. 42, No. 6A, (2003), pp. 3666-3675.

Yokoyama, T.; Ishii, H.; Matsuie, N.; Kanai, K.; Ito, E.; Fujimori, A.; Araki, T.; Ouchi, Y.; Seki, K., Neat Alq_3 thin film and metal/Alq_3 interfaces studied by NEXAFS spectroscopy, *Synth. Met.* Vol. 152, No. 1-3, (2005), pp. 277-280.

Yoshizaki, K.; Manaka, T.; Iwamoto, M., Large surface potential of Alq_3 film and its decay, *J. Appl. Phys.*, Vol. 97, No. 2, (2005), pp. 023703.

Zhu, R.; Lin, J. M.; Wang, W. Z.; Zheng, C.; Wei, W.; Huang, W.; Xu, Y. H.; Peng, J. B.; Cao,Y., Use of the β-phase of poly(9,9-dioctylfluorene) as a probe into the interfacial interplay for the mixed bilayer films formed by sequential spin-coating, *J. Phys. Chem. B*, Vol. 112, No. 6 (2008), pp. 1611-1618.

Zhuang, X.; Miranda, P. B.; Kim, D.; Shen, Y. R., Mapping molecular orientation and
    conformation at interfaces by surface nonlinear optics, *Phys. Rev. B*, Vol. 59, No. 19
    (1999), pp. 12632-12640.

7

Low-Frequency Coherent Raman Spectroscopy Using Spectral-Focusing of Chirped Laser Pulses

Masahiko Tani[1], Masakazu Hibi[1], Kohji Yamamoto[1], Mariko Yamaguchi[2],
Elmer S. Estacio[1], Christopher T. Que[1] and Masanori Hangyo[3]
[1]*Research Center for Development of Far-Infrared Region, University of Fukui,*
[2]*Graduate School of Materials Science, Nara Institute of Science and Technology,*
[3]*Institute of Laser Engineering, Osaka University*
Japan

1. Introduction

Vibrational spectroscopy is generally implemented using two schemes; that is, absorption spectroscopy and Raman spectroscopy. Conventionally, low frequency absorption spectroscopy is carried out using Fourier transform spectrometers equipped with a far infrared radiation source and a thermal detector. On the other hand, low frequency Raman spectroscopy is carried out by way of double or triple monochromaters and high-quality notch filters, whose performance determines the low frequency limit of the Raman spectrometer. In addition, in recent years, terahertz time-domain spectroscopy (THz-TDS) (Hangyo et al., 2005), utilizing femtosecond lasers as the excitation source, has been developed. THz-TDS enabled us to obtain absorption and dispersion spectra with a high signal-to-noise ratio in a frequency region less than 3 THz (100 cm⁻¹) and can be applied for absorption spectroscopy of various substances (Kawase et al., 2009; Korter et al., 2006; Taday et al., 2003;Tani et al., 2004; Tani et al., 2010; Walther et al., 2003; Yamamoto et al., 2005; Yamaguchi et al., 2005) and imaging measurements (Kawase, 2004).

There is a keen interest in low frequency vibrational spectroscopy for biomolecules since large amplitude, low frequency modes in macro biomolecules are believed to be associated with their respective function, as in the case of proteins (Chou, 1985, 1988). In order to fully understand the dynamics and function mechanisms of biomolecules, it is necessary to study their large amplitude and anharmonic low-frequency vibrational motions as these govern their thermal and physiochemical properties. A normal mode analysis of protein molecules revealed that large amplitude vibrational modes which are delocalized in the whole molecule lie within the THz region (< 120 cm⁻¹) (Brooks & Karplus, 1985; Go et al., 1983). In addition, the calculations suggested that the entropy of the whole molecule, specifically its thermodynamic characteristics, is governed by the large amplitude vibrational modes in the sub-THz region (<30 cm⁻¹). Consequently, important information related to the functions and dynamics of proteins can be derived by investigating THz vibration spectra.

For observation of low frequency vibrational modes in macro biomolecules, the absorption spectroscopy, including THz-TDS, has not been very successful because of the strong absorption of water in aqueous or hydrated samples. Low frequency Raman spectroscopy, on the other hand, seems to be promising for the study of low frequency vibrational modes in macro biomolecules since Raman spectroscopy is less influenced by water molecules as compared to THz absorption spectroscopy. As such, there have been some reports on the use of low frequency Raman spectroscopy to observe low frequency vibrational modes in proteins, such as lysozyme (Genzel et al., 1976; Urabe et al., 1998).

Some difficulties arise in carrying out the low frequency Raman or low frequency coherent Raman spectroscopy, in contrast to the high frequency regime; especially in the case of biological samples. One such problem is the strong Rayleigh scattering from rugged surfaces like powdered samples and from large biomolecules such as proteins and DNA. Therefore, to obtain a clear Raman spectrum, an efficient notch filter, such as the vapour iodine filter (Okajima & Hamaguchi, 2009), is required. In addition, slow signal fluctuations arising from thermal gradients due to laser beam heating also poses problems. To remove the thermal fluctuation, a noise subtraction technique using a lock-in amplifier has been reported (Genzel et al., 1976).

A more significant drawback in using Raman Spectroscopy, however, is its inherently low signal intensity. To overcome the low signal intensity in spontaneous Raman spectroscopy, coherent Raman techniques, such as Coherent Anti-Stokes Raman Scattering (CARS), may be utilized. Using the coherent Raman scattering, the signal can be increased by 5 to 6 orders of magnitude compared to that of spontaneous Raman scattering. With this high signal intensity, CARS from the C-H stretching mode has been successfully used for the molecular imaging of biological cells and tissues.

In line with this, the authors have recently developed coherent Raman spectroscopy technique in the THz frequency region aimed for imaging and spectroscopy of biomolecules. This technique uses a broadband femtosecond laser as the light source, as opposed to the customary nano- to picosecond monochromatic laser, in exciting the coherent Raman scattering and in detecting the signals in the time-domain. The femtosecond laser is advantageous since it can also be used in a THz-TDS system. At the moment, this technique is still being fully developed and has not been applied in the spectroscopy of biomolecules nor in the spectroscopic imaging of living tissues. The authors discuss the concept of the time-domain coherent Raman spectroscopy based on the "spectral focusing" technique. Furthermore, we demonstrate a series of proof-of-principle measurements by using a semiconductor sample. Observations on the optical phonon bands of GaSe in the THz frequency through the time-domain CARS and inverse Raman spectroscopy are presented. Lastly, efforts on the improvements in the signal-to-noise ratio (SNR) for future measurements of biomolecules are discussed.

2. Principle of the coherent Raman spectroscopy in terahertz frequency region using spectral focusing of femtosecond laser

2.1 Basics of coherent Raman spectroscopy

The basic principle of Raman scattering is explained briefly prior to discussing coherent Raman spectroscopy in the terahertz frequency region. A polarization $P^{(1)}$ is induced when a molecule or crystal lattice is placed under an electric field, E_0. The induced polarization depends linearly on the applied electric field through the polarizability α when the field is not so strong, and it can be described by

$$P^{(1)} = \alpha\, E_0. \tag{1}$$

In the above equation, the polarization $P^{(1)}$ represents the polarization of one molecule, or the induced macroscopic polarization by the molecules or crystal lattice per unit volume. Additionally, the polarizability α is dependent on the molecule; hence any periodic change in the molecule from thermally excited molecular or lattice vibration, will also periodically change α. When a laser of frequency ω_0 with an electric field amplitude E_0 is incident onto a material, wherein its constituent molecules fluctuate periodically at a vibration frequency $\delta\omega$, the polarization $P^{(1)}$ will oscillate at the same frequency ω_0 of the incident laser and, at the same time, $P^{(1)}$ will be modulated by the change in α induced by molecular vibrations. This will then lead to the sum frequency $\omega_0 + \delta\omega$ and difference frequency $\omega_0 - \delta\omega$. Specifically, the polarization $P^{(1)}$ will oscillate with the three frequency components ω_0, $\omega_0 + \delta\omega$, and $\omega_0 - \delta\omega$. Since the emitted electromagnetic wave is proportional to the change in the electric polarization $\partial^2 P^{(1)} / \partial t^2$, light will be scattered by the molecule according to three mechanisms: (1) light with the same frequency as the incident light ω_0 (Rayleigh scattering), (2) positive shift $\omega_0 + \delta\omega$ with the vibration frequency $\delta\omega$ (anti-Stokes), and (3) negative shift $\omega_0 - \delta\omega$ with the vibration frequency $\delta\omega$ (Stokes). For thermally excited molecular vibrations, the vibration amplitude and phase is small and incoherent, respectively. Thus, the positive and negative interference of the emission from each molecule will occur with equal probabilities which prevents a large scattering intensity to be obtained. Also, the Stokes and anti-Stokes signals are relatively weak compared to the Rayleigh scattering. In this regard, an efficient notch filter and monochromator are needed in order to suppress the Rayleigh scattering. On top of this, a highly sensitive photomultiplier tube or CCD camera are needed to observe the weak Stokes and anti-Stokes signals. Additionally, it should be noted that the polarization $P^{(1)}$ is a vector and since the polarization is dependent on the vector component of the incident electric field, the polarizability α is a tensor (Raman tensor). However, we will not consider its tensor properties at this time, for simplicity.

Now, let us consider a situation that two electric fields, E_1 and E_2 with frequencies ω_1 and ω_2, respectively are incident on the material in addition to E_0. Three-wave mixing occurs with these two waves and E_0, to induce a polarization $P^{(3)}$. In this case we have

$$P^{(1)} = \alpha\, E_0 E_1 E_2. \tag{2}$$

This nonlinear optical interaction is in fact a "four-wave mixing" process since there are four electro-magnetic waves involved, which includes the wave generated by the nonlinear polarization $P^{(3)}$ in addition to the three incident waves. When the difference frequency of E_1 and E_2 is in resonance with the molecular vibration frequency $\delta\omega$ (that is, $\delta\omega = |\,\omega_1 - \omega_2|$) the resonance effect will cause the molecule to oscillate coherently such that the polarization $P^{(3)}$ will vary significantly with the vibration frequency $\delta\omega$. During this process, three frequency components ω_0, $\omega_0 + \delta\omega$, and $\omega_0 - \delta\omega$ will compose the scattered light. In this case, the scattered light with frequency $\omega_0 + \delta\omega$ or $\omega_0 - \delta\omega$ is enhanced by the coherence effect (Note that with N numbers of molecules emitting coherently, the increase in the emission intensity is not N times but rather N^2 times in intensity). The scattered light with frequency $\omega_0 + \delta\omega$ is called Coherent Anti-Stokes Raman Scattering (CARS) while the scattered light with frequency $\omega_0 - \delta\omega$ is called Coherent Stokes Raman Scattering (CSRS). Given that it is not easy to have a laser source with three different frequencies, the frequency of the field E_0 usually coincides with either E_1 or E_2 (ω_1 or ω_2). Taking into account the phase of each light wave, Eq. (2) is rewritten as,

$$CARS : P^{(3)}\,(\omega_1 + \delta\omega) = \alpha E_1\,(\omega_1)E_1\,(\omega_1)E_2{}^*\,(\omega_2)\quad(\omega_1 + \delta\omega = \omega_1 + \omega_1 - \omega_2\,,\;\omega_1 > \omega_2) \qquad (3a)$$

$$CSRS : P^{(3)}\,(\omega_2 - \delta\omega) = \alpha E_1{}^*\,(\omega_1)E_2\,(\omega_2)\,E_2\,(\omega_2)\quad(\omega_1 - \delta\omega = \omega_2 + \omega_2 - \omega_1\,,\;\omega_1 > \omega_2) \qquad (3b)$$

Here, E^* is the phase conjugate of E.

In Eq. (3a), the interaction of the photons of ω_1 and ω_2 with the molecule (lattice vibration) leads to the generation of a photon having frequency $\omega_1 + \delta\omega$ through the 3rd order nonlinear optical process. In the CSRS process shown in Eq. (3b), the interaction of the photons of ω_1 and ω_2 with the lattice vibration leads to the generation of a photon with frequency $\omega_2 - \delta\omega$ via the same optical process. In addition to the CARS and CSRS, four-wave mixing also leads to inverse Raman scattering (IRS), which is the annihilation of a ω_1 ($= \omega_1 - \omega_2 + \omega_2$) photon; or stimulated Raman gain scattering (SRGS), which is the creation of a ω_2 ($= \omega_2 + \omega_1 - \omega_1$) photon. The nonlinear polarization corresponding to each of these scattering processes are given by the following equations:

$$IRS : P^{(3)}\,(-\omega_1) = \alpha E_1{}^*\,(\omega_1)E_2\,(\omega_2)E_2{}^*\,(\omega_2)\quad(-\omega_1 = -\omega_1 + \omega_2 - \omega_2\,,\;\omega_1 > \omega_2) \qquad (3c)$$

$$SRGS : P^{(3)}\,(\omega_2) = \alpha E_2\,(\omega_2)E_1\,(\omega_1)E_1{}^*\,(\omega_1)\quad(\omega_2 = \omega_2 + \omega_1 - \omega_1\,,\;\omega_1 > \omega_2). \qquad (3d)$$

The IRS and SRGS processes represent two aspects of the energy exchange between the incident fields E_1 and E_2: In IRS we observe a reduction of intensity in E_1 while in SRGS we observe an increase of E_2 as a result of energy transfer from E_1 to E_2 through the coherent Raman process.

2.2 Principle of time-domain coherent Raman spectroscopy based on spectral focusing

For a coherent Raman spectroscopy system that uses two excitation light sources ω_1 and ω_2, two frequency-stabilized and frequency-tunable lasers are needed. This will make the system bulky. Moreover, the excitation lasers should have a narrow spectral linewidth (< 1 cm-1), and a very efficient notch filter, having sufficient optical density and a sharp spectral edge, are required to filter out the Rayleigh scattering. To overcome this problem, a technique using the chirped frequency of the femtosecond laser was developed. Although the details of this technique will be explained below using CARS as an example, this technique is also applicable to other coherent Raman processes.

A femtosecond laser has a broad spectral bandwidth extending from several THz to a few tens of THz. The frequency of the femtosecond laser pulse can be chirped using a grating pair. After passing through the gratings, the chirped laser pulse is separated into two to produce Pump1 (E_1) and Pump2 (E_2) and combined again in an interferometer, where one pump pulse passes through an optical delay line. If the two pump pulses overlap with a time difference $\Delta\tau$, a $\Delta\tau$-dependent optical beat arising from the beat frequency will be generated. If this optical beat is incident onto a sample material, a frequency-chirped and up-converted CARS signal can be obtained, as seen on the left illustration in Fig.1. This technique for obtaining a narrow band signal from broadband, chirped pump pulses is

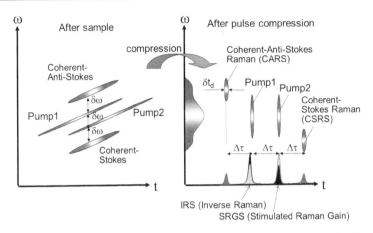

Fig. 1. Frequency chirped Pump1, Pump2, and coherent Raman emissions (Left), and those of after pulse compression (Right) in the time-frequency plane.

called "spectral focusing", reported by Hellerer *et al* in 2004 (Hellerer et al., 2004). Using a spectral focusing technique with a femtosecond laser, they successfully observed C–N Raman bands of nitroprusside [$Fe(CN)_5NO^{2-}$] at around 2160 cm^{-1} with three sharp peaks separated ~15 cm^{-1} apart from each other.

However, if the difference frequency, $\delta\omega$, in the CARS signal is small, the signal, Pump1 and Pump2 beams are spectrally overlapped and cannot be separated by dispersing these spectra as is usually done in Raman spectrum measurements. In addition, the CARS signal beam is almost collinear with the pump beams due to the phase-matching condition in the THz regime.

$$\mathbf{k}_{CARS}=2\mathbf{k}_1-\mathbf{k}_2\cong\mathbf{k}_1(\cong\mathbf{k}_2) \qquad (4)$$

Here $\mathbf{k}_{CARS}$ is the wave vector of the CARS signal and $\mathbf{k}_1$ ($\mathbf{k}_2$) is that of Pump1 (Pump2). They can be separated only in time domain by pulse compression through the inverse chirping process, as shown by the right illustration in Fig. 1. The detection of the CARS signal in time-domain is possible using the up-conversion technique. Only one femtosecond laser is needed with this technique and neither a monochromater nor a filter is necessary. Likewise, it is possible to detect the CSRS signal of Pump2 at a frequency $\omega_2-\delta\omega$. Lastly, by means of the above technique, the IRS and SRGS signal can be detected by monitoring intensity changes in the coherent Raman scattering. The time-domain coherent Raman process resembles the chirped pulse amplification (CPA) technique, with which the frequency chirped pulse is amplified and compressed to obtain a high energy femtosecond laser pulses.

In time-domain coherent Raman spectroscopy, the frequency bandwidth of the detected coherent Raman signal is limited by the spectral bandwidth of the pump beams since the signal is the result of the difference frequency mixing of the two pump beams. On the other hand, the frequency resolution is determined by the reciprocal of the time-width ΔT of the two chirped optical pulses. The frequency resolution Δv is given by

$$\Delta v (= \Delta\omega/2\pi)=1/\Delta T. \qquad (5)$$

The time-width of the optical beat is maximum when the associated beat frequency (Raman frequency) is v~0, and decreases proportionately with the increase of the beat frequency. Therefore, the frequency resolution Δv increases as the Raman frequency decreases. The highest frequency resolution obtainable at the low frequency limit for a femtosecond laser with a transform-limited pulse width of δt and a chirp rate b is given by

$$\Delta v_m = 1/ \Delta T_m = \delta \tau \cdot b, \qquad (6)$$

where ΔT_m is the pulse duration of the chirped pump pulse. Accordingly, in order to get a good frequency resolution, one should use a laser with a narrow pulse width and be able to significantly stretch the pulse.

2.3 Experimental setup for time-domain coherent Raman spectroscopy

The schematic illustration of the experimental setup is shown in Fig. 2. Femtosecond pulses from a Ti: sapphire regenerative amplifier system (λ ~800 nm, δt ~120 fs, 1 kHz, 800 µJ/pulse) were initially split into pump and probe pulses. The pump pulse was positively chirped with a grating-lens pair and stretched to about 30 ps. Using a Michelson-type interferometer, the stretched pump pulse was divided into two beams, Pump1 and Pump2, after which the two beams acquired a relative time-delay, $\Delta \tau$, and were then recombined. The interference of the two pump beams produced an optical beat at the beams' instantaneous difference frequency. The interfering pump beams were then directed to the sample material. Subsequently, the pump beams were compressed to their original fs pulse widths using another grating pair. The frequency chirped CARS signal was also compressed and separated from Pump1 by $\Delta \tau$ in time domain. The CARS signal was then up-converted to ~400-nm wavelength through sum frequency generation (SFG) with the probe pulse using a BBO crystal. The up-converted signal was then detected by a GaP photodiode. In order to have a high SNR, either Pump1 or Pump2 should be modulated by a mechanical chopper and the photodiode signal can then be detected by a lock-in amplifier.

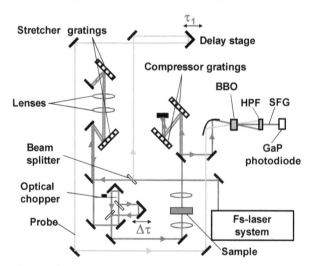

Fig. 2. Experimental setup for time-domain coherent Raman spectroscopy

Figure 3(a) shows the SHG cross-correlation signal for stretched Pump1 and Pump2 with a time difference, whose beat frequency is 736 GHz. Figure 3(b) shows the Fourier transformed SHG cross-correlation signal for stretched Pump1 and Pump2 with various $\Delta\tau$ values. By plotting the observed beat frequencies against the optical delay $\Delta\tau$ we can determine the chirp rate b of the pump pulses. From Fig. 3(b) it is also found that the spectral line width of the optical beat increases with increasing beat frequency, resulting in poor spectral resolution of the time-domain coherent Raman spectra at higher frequencies. The frequency resolution, $\Delta\nu_m$, in the low frequency limit ($\nu\sim 0$) is calculated to be 0.02 THz for the present system with the pulse width δt = 120 fs and the chirp rate b =0.18 THz/ps. The useful bandwidth of the CARS measurement system estimated by the optical beat measurement is about 5 THz. The low-frequency limit of the CARS measurement is determined by the pulse width of the pump laser after the compression and is about 0.2 THz in the present system.

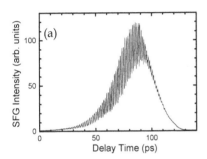
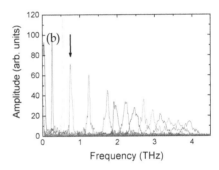

Fig. 3. (a) SHG cross-correlation signal of stretched Pump1 and Pump2. (b) Fourier transformed cross-correlation signal of stretched Pump1 and Pump2 with various $\Delta\tau$. The Fourier transformed spectrum of SHG correlation signal shown in (a) is indicated by a vertical arrow.

3. Time-domain coherent Raman spectroscopy

3.1 Time-domain coherent anti-Stokes Raman spectroscopy (CARS) for GaSe

Figure 4 shows the CARS spectrum measured from a β-GaSe single crystal (c-cut, 1 mm thickness) sample (Tani et al., 2010). The spectrum shown in Fig. 4 was obtained by "peak-scanning", where the time-delay τ_1 for the probe and the relative time difference, $\Delta\tau$, between Pump1 and Pump2 are simultaneously scanned while the ratio of the two delays, $\tau_1/\Delta\tau$, was kept equal to 2 (τ_1=2 $\Delta\tau$). In effect, the frequency shift of the CARS signal with respect to the change of the difference frequency $\delta\omega$ was investigated. To increase the SNR the "peak-scanned" spectrum was averaged for 10 times.

On the broad, non-resonant background spectrum a resonance can be seen near 0.6 THz (~20 cm^{-1}), coming from the lowest Raman-active optical mode (E$_{2g}$ mode) in β-GaSe (Wieting & Verble, 1972). The non-resonant coherent Raman signal originates from electronic response of the sample and is not strongly frequency dependent. The resonant CARS spectrum is not symmetric but has a dispersion-type structure. This is explained by the interference between the components from the non-resonant (real and constant with frequency) and the resonant nonlinear optical susceptibility (Levenson, 1974). The CARS signal intensity is proportional to

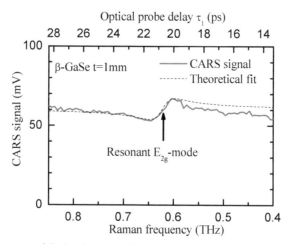

Fig. 4. CARS spectrum of GaSe showing the optical phonon band near 0.6THz

$$| \chi^{(3)} |^2 = | \chi_N^{(3)} + \chi_R^{(3)} |^2$$

$$= | \chi_N^{(3)} |^2 + \chi_N^{(3)} \frac{a(\omega_d - \omega_R)}{(\omega_d - \omega_R)^2 + \Gamma^2} + \frac{a^2}{(\omega_d - \omega_R)^2 + \Gamma^2} \qquad (7)$$

Here, $\chi_N^{(3)}$ is the non-resonant third order susceptibility, which is assumed to be a real constant, while the resonant third order susceptibility $\chi_R^{(3)}$ is given in the following form:

$$\chi_R^{(3)}(\omega_d = \omega_p - \omega_S) = \frac{a(\omega_d - \omega_R + i\Gamma)}{(\omega_d - \omega_R)^2 + \Gamma^2} \qquad (8)$$

Here, $\omega_d = \omega_p - \omega_S > 0$ is the difference frequency between the pump (Pump1) and Stokes (Pump2), ω_R is the Raman resonance frequency, Γ is the Raman linewidth, and a is the constant associated with the magnitude of the resonance. When the non-resonant contribution is large compared to the resonant one, we can neglect the last term in Eq.(7). Then we obtain

$$\text{CARS} \propto \ | \chi_N^{(3)} |^2 + 2\chi_N^{(3)} \frac{a(\omega_d - \omega_R)}{(\omega_d - \omega_R)^2 + \Gamma^2} = | \chi_N^{(3)} |^2 + 2\chi_N^{(3)} \frac{a\Delta\omega}{\Delta\omega^2 + \Gamma^2} \qquad (9)$$

Here, $\Delta\omega = \omega_d - \omega_R$ is the detuning from the resonant Raman frequency. The schematic illustration of the CARS signal interfering with the resonant and non-resonant components is shown in Fig. 5 The curve fit using eq.(3) is shown in Fig. 4 as the dashed line. We have estimated the Raman linewidth $\Gamma = 0.02$ THz (=0.67 cm^{-1}) and the ratio of the resonant to the non-resonant contribution, given by the parameter $a / (\Gamma \chi_N^{(3)})$, to be 12%. The discrepancy between the theoretical curve fit and the observed spectrum in the low-frequency side is due to the drift of the non-resonant background signal caused by the fluctuation of the laser intensity.

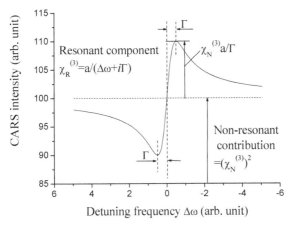

Fig. 5. The structure of CARS spectrum resulting from interference between the non-resonant and resonant components

3.2 Time-domain inverse Raman spectroscopy (IRS) for GaSe

Figure 6 shows the inverse Raman spectrum of the β–GaSe sample. The total pump power incident on the sample was about 1 mW. The signal loss of Pump1 was detected using lock-in techniques by modulating Pump2 with a mechanical chopper. The IRS spectrum was taken by scanning the Pump2, $\Delta\tau$, with the probe optical delay set at the maximum SFG signal . As can be seen from Fig. 6, there are resonance peaks around 0.6 and 4.2 THz due to the E_{2g} and A_{1g} optical phonon modes, respectively. The resonance peak expected at 1.8 THz corresponding to an optical phonon (E_{1g}-mode) in GaSe was not observed, probably due to the weak Raman scattering cross section. The inverse Raman signal is one order stronger than the CARS signal. Accordingly, the SNR of the inverse Raman signal is also larger. The reason for this is not clear at present.

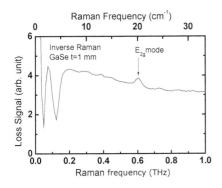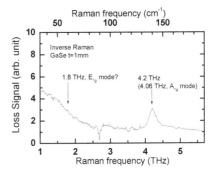

Fig. 6. Inverse Raman spectra of GaSe

One of the advantages of the time-domain IRS (and SRGS) in comparison with the time-domain CARS (or CSRS) is that it is not necessary to scan the optical delay τ_1 of the probe light simultaneously with $\Delta\tau$. In addition, the IRS spectrum is symmetric about the resonance frequency while that of CARS is not symmetric (dispersive shape). This is because the signal in an IRS spectrum corresponds to the imaginary part of the 3rd order nonlinear susceptibility (the same as the spontaneous Raman signal) while that of CARS corresponds to the real part, and this is the reason why it interferes with the signal from the real non-resonant nonlinear susceptibility. The non-resonant background signal is also dominant as in CARS and is probably due to other non-resonant four-wave mixing processes. Therefore, it is necessary to suppress this non-resonant background signal when performing spectroscopy of biomolecules and imaging where a higher SNR is required.

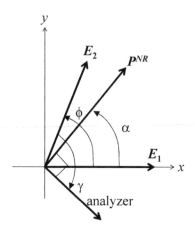

Fig. 7. Polarization vectors of Pump1 (E_1) and Pump2 (E_2), the non-resonant IRS signal (P^{NR}) and the analyser polarizer.

The polarized CARS (P-CARS) technique (Oudar et al., 1979; Cheng et al., 2001) can be applied to the inverse Raman measurement in order to suppress the non-resonant background signal. This is achieved by controlling the relative polarization angle between Pump1 and Pump2; that is, adjusting the polarization direction of the non-resonant and resonant signals. Owing to the polarization difference, it is possible to remove the polarization component of the non-resonant signal by placing a polarizing filter in front of the detector. The principle of polarization IRS method (P-IRS) is outlined as follows. Consider that Pump1 is linearly polarized along the x-axis and Pump2 is polarized along a direction, an angle ϕ relative to the x-axis as shown Fig.7. When the beat frequency generated by the two pump beams is resonant with a molecular vibration or an optical phonon vibration, the interaction of the pump beam in the sample induces a third order nonlinear polarization that contains a non-resonant part, P^{NR}, and a resonant part, P^R. The x and y components of the non-resonant part contributing to IRS can be written as

$$P_x^{NR} = 3\chi_{1111}^{NR} E_1{}^* E_2\, E_2{}^* \cos^2\phi,$$

$$P_y^{NR} = 3\chi_{2112}^{NR} E_1{}^* E_2\, E_2{}^* \cos\phi\sin\phi \qquad (10)$$

Similarly, the x and y components of the resonant part can be written as

$$P_x^R = 3\chi_{1111}^R E_1^* E_2 \, E_2^* \cos^2 \phi \, ,$$

$$P_y^R = 3\chi_{2112}^R E_1^* E_2 \, E_2^* \cos\phi\sin\phi \tag{11}$$

In the absence of electronic resonance, χ^{NR} is a real quantity and independent of frequency. In this case the depolarization ratio of the non-resonant IRS field is

$$\rho^{NR} = \frac{\chi_{2112}^{NR}}{\chi_{1111}^{NR}} = \frac{1}{3} \tag{12}$$

P^{NR} is therefore linearly polarized with an angle α relative to the x axis,

$$P^{NR} = 3\chi_{1111}^{NR} E_1^* E_2 \, E_2^* \cos^2 \phi / \cos\alpha \tag{13}$$

where the angle α is related to ϕ by $\tan\alpha = \rho^{NR} \tan\phi$.
The non-resonant background can be removed by placing an analyzer in front of the detector, with its polarization perpendicular to P^{NR}. The total projection of the two components of P^R [Eq. (11)] along the direction perpendicular to P^{NR} can be written as

$$P_\perp = 3\chi_{1111}^R E_1^* E_2 \, E_2^* (\cos^2\phi\sin\alpha - \rho^R \cos\phi\sin\phi\cos\alpha) \tag{14}$$

Here,

$$\rho^R = \frac{\chi_{2112}^R}{\chi_{1111}^R} \tag{15}$$

is the depolarization ratio of the resonant IRS field, and is equal to the spontaneous Raman depolarization ratio in the absence of electronic resonance.

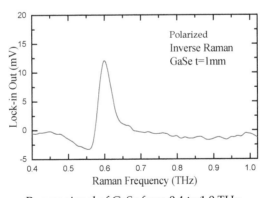

Fig. 8. Polarized inverse Raman signal of GaSe from 0.4 to 1.0 THz

Even as this technique causes a slight reduction of the resonant signal, a large amount of the non-resonant signal is suppressed. Figure 8 shows the inverse Raman spectrum using the polarization technique. As can be seen from the figure, the non-resonant component is well suppressed and the resonance peak is clearly seen at the optical phonon mode frequency around 0.6 THz.

4. Extension of measurement bandwidth

The measurement bandwidth is dependent on the spectral bandwidth and thus the associated pulse width of the femtosecond laser. Thus, using a narrow pulse width femtosecond laser will give a broad measurement bandwidth. However, when using very short pulses (less than 100 fs), special care must be taken to suppress higher order dispersion in the stretcher configuration. With a grating pair such as the one shown in Fig.2 (the compressor part), optical pulses with negative chirp can be obtained. When a lens pair is inserted between the grating pair, the spatial image of the optical beam is inverted and a negative chirp is then reverted to a positive chirp. However, the insertion of a lens pair introduces higher order dispersion and spherical aberration, which cannot be compensated by the grating pair compressor. As a result, the pulse width after compression is broadened with the higher order dispersion and the time-resolution in the detection system deteriorates. To avoid dispersive optics in the stretcher system we adopted Öffner configuration (Öffner, 1971; Cheriaux et al., 1996), which consists of a grating and a combination of concave, convex and roof mirrors as shown in Fig. 9. The optical pulses hit the grating and the aspheric mirrors four times. By changing the offset distance Z_1 of the grating position we can control the positive dispersion of the optical pulses.

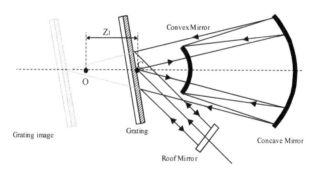

Fig. 9. The Öffner configuration used as the stretcher. The point O is the center of curvature of both mirrors. The point C is the incident position of the optical pulses on the grating. The point C lies on the incidence plane of the optical pulses on the grating.

Figure 10 shows the measured inverse Raman spectrum from the β–GaSe using femtosecond laser pulses with a pulse width of 40 fs in the transform-limited condition. The vertical axis of the graph is the optical power loss in Pump1, corresponding to the inverse Raman signal, while the horizontal axis is the relative time-delay between the two pump pulses, which determines the beat frequency. The sharp peak at the zero time-delay is the burst signal of the overlapping pump pulses, which corresponds to zero-beat frequency. The small and sharp peak at 1.1 ps (~0.6 THz) is the optical phonon mode (E_{2g} mode), corresponding sliding motion of layers with its plane perpendicular to the c-axis of GaSe. The apparent peak at 7.9 ps (4 THz) is the fully symmetric phonon mode (A_{1g} mode) and the peak at 18.7 ps (9.3 THz) is another Raman active mode (A_{1g} mode) (Wieting & Verble, 1972). Based on these bands and the associated non-resonant background signal, the frequency bandwidth of this system is estimated to be 15 THz (corresponding to ~30 ps time delay).

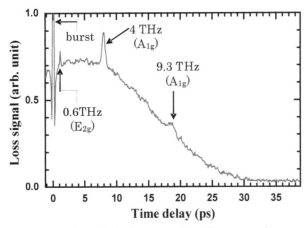

Fig. 10. Inverse Raman signal from GaSe (c-cut, t = 1 mm) measured using a 40 fs pulse width laser

5. Problem and future plan

A crucial aspect of the coherent Raman spectroscopy system is its utilization of a femtosecond laser in conjunction with a novel time-domain detection scheme to acquire the sub-THz to THz coherent Raman spectrum. However, low signal strength from samples like biomolecules necessitates improving the SNR of the system. The efficiency of coherent Raman scattering and the SFG process in the signal detection is proportional to the cube and square of the excitation intensity, respectively. Therefore, the coherent Raman scattering signal detected in time-domain is proportional to the 4th power of the excitation intensity. Laser intensity fluctuations in turn, cause large fluctuations in the signal. This deteriorates the SNR, even if the signal itself is large. Moreover, the large fluctuation of the non-resonant signal compared to the resonant signal due to vibrational modes also results in a low SNR. In order to address these issues, the following points may be considered: (1) use a high repetition rate laser, (2) use balanced detection or obtain the difference between the signal and the reference, and (3) establish ways of measuring the differential signal in the time delay of Pump1 and Pump2 (equivalent to the differential signal in the frequency domain) through the modulation of their optical path lengths. If the repetition rate of the pump laser is too high, as in the case of a mode-locked Ti:sapphire laser oscillator (having typical repetition frequencies of 50~100 MHz), the energy per pulse is too small and the coherent Raman signal strength is too weak resulting to a low SNR. For this reason, the optimum pump source might be a femtosecond laser amplifier system with a repetition rate of ~100-kHz and with an energy per pulse of a few µJ, in order to maintain sufficient excitation intensity.

6. Conclusion

THz time-domain coherent Raman spectroscopy system using "spectral focusing" of a broadband femtosecond laser source was introduced. Although the system is still in its development stage, sufficient SNR was obtained using a GaSe sample. Additionally, it promises to be a reliable measurement technique for background light and fluorescence studies since the signal is gated in the time domain. Lastly, improving the SNR characteristics of the system will make it feasible for applications in the spectroscopy and imaging of biomolecular samples.

7. Acknowledgment

The Authors are grateful to the "SAKIGAKE (PREST)" Grant in the research area "Life Phenomena and Measurement Analysis" by the Japan Science and Technology Agency, and the Grants-in-Aid for Scientific Research (KAKENHI) (B) program by the Japan Society for the Promotion of Science (JSPS).

8. References

Brooks, B. & Karplus, M. (1985). Normal Modes for Specific Motions of Macromolecules: Application to the Hinge-Bending Mode of Lysozyme. *Proceedings National Academy of Sciences*, Vol.82, No.15, (August 1985), pp. 4995-4999

Cheng, J.-X.; Book, L. & Xie, X. S. (2001). Polarization Coherent Anti-Stokes Raman Scattering Microscopy. *Optics Letters*, Vol.26, No.17, (September 2001), pp. 1341-1343

Cheriaux, G.; Rousseau, P.; Salin, F.; Chambaret, J. P.; Walker, B. & L. F. Dimauro. (1996) Aberration-free Stretcher Design for Ultrashort-pulse Amplification. *Optics Letters*, Vol.21, No. 16, (March 1996), pp. 414-416

Chou, K.-C. (1985) Low-frequency Motions in Protein Molecules. *Journal of Biophysics*, Vol.48, No.2, (August 1985), pp. 289-297

Chou, K.-C. (1988) Low-frequency Collective Motion in Biomacromolecules and its Biological Functions. *Biophysical Chemistry*, Vol.30, No.1, (May 1988), pp. 3-48

Genzel, L.; Keilmann, F.; Martin, T. P.; Winterling, G.; Yacoby, Y.; Frohlich, H. & Makinen, M. (1976). Low-Frequency Raman Spectra of Lysozyme. *Biopolymers*, Vol.15, No.1, (January 1976), pp. 219-225

Go, N.; Noguti, T. & Nishikawa, T. (1983). Dynamics of A Small Globular Protein in Terms of Low-frequency Vibrational Modes. *Proceedings National Academy of Sciences*, Vol.80, No. 12, (June 1983), pp. 3696-3700

Hangyo, M.; Tani, M. & Nagashima, T. (2005). Terahertz Time-Domain Spectroscopy of Solids: A Review. *International Journal of Infrared and Millimeter Waves*, Vol.26, No.12, (December 2005), pp. 1661-1690

Hellerer, T.; Enejder, A. M. K. & Zumbusch, A. (2004). Spectral Focusing: High Spectral Resolution Spectroscopy with Broad-bandwidth Laser Pulses. *Applied Physics Letters*, Vol.85, No.25, (March 2004), pp. 25-27

Kawase, K. (2004). Terahertz Imaging For Drug Detection and Large-scale Integrated Circuit Inspection. *Optics & Photonics News*, Vol.15, No.10, (October 2004), pp. 34-39

Kawase, M.; Saito, T.; Ogawa, M.; Uejima, H.; Hatsuda, Y.; Kawanishi, S.; Hirotani, Y.; Myotoku, M.; Ikeda, K.; Takano, K.; Hangyo, M.; Yamamoto, K. & Tani, M. (2009).

Terahertz Absorption Spectra of Original and Generic Ceftazidime. *Analytical Sciences*, Vol.25, No.12, (November 2009), pp. 1483-1485

Korter, T. M.; Balu, R.; Campbell, M. B.; Beard, M. C.; Gregurick, S. K. & Heilweil E. J. (2006). Terahertz Spectroscopy of Solid Serine and Cysteine. *Chemical Physics Letters*, Vol.418, No.1, (November 2005), pp. 65-70

Levenson, M. C. (1974). Feasibility of Measuring the Nonlinear Index of Refraction by Third-order Frequency Mixing. *IEEE Journal of Quantum Electronics*, Vol.10, No.2, (February 1974), pp. 110-115

Öffner, A. U.S. patent 3,748,015 (1971)

Okajima, H. & Hamaguchi, H. (2009). Fast Low Frequency (Down to 10 cm^{-1}) Multichannel Raman Spectroscopy Using an Iodine Vapor Filter. *Applied Spectroscopy*, Vol.63, No.8 (August 2009), pp. 958-960

Oudar, J. L.; Smith, R. W. & Shen, Y. R. (1979). Polarization-sensitive Coherent Anti-Stokes Raman Spectroscopy. *Applied Physics Letters*, Vol.34, No.11, (June 1979), pp. 758-760

Taday, P. F.; Bradley, I. V.; Arnone, D. D. & Pepper, M. (2003). Using Terahertz Pulse Spectroscopy to Study the Crystalline Structure of a Drug: A Case Study of the Polymorphs of Ranitidine Hydrochloride. *Journal of Pharmaceutical Science*, Vol.92, No.4, (April 2003), pp. 831-838

Tani, M.; Yamaguchi, M.; Miyamaru, F.; Yamamoto, K. & Hangyo, M. (2004). Spectroscopy of Biomolecules using Terahertz Electromagnetic Pulse–Measurement of the Low Vibration Mode of Amino Acids using Terahertz Time-domain Spectroscopy. *Optical Alliance*, Vol.15, No.1, (2004), pp.9-14 (in Japanese)

Tani, M.; Koizumi, T.; Sumikura, H.; Yamaguchi, M.; Yamamoto, K. & Hangyo, M. (2010). Time-Domain Coherent Anti-Stokes Raman Scattering Signal Detection for Terahertz Vibrational Spectroscopy Using Chirped Femtosecond Pulses. *Applied Physics Express*, Vol.3, (July 2010), pp. 072401-1-072401-3

Tani, M. & Hangyo, M. (2010). Terahertz Wave Emission and Application using a Femtosecond Solid-state Laser in the Special Issue on Wavelength Conversion Technology with Solid-state Lasers and its Applications. *Optical Alliance*, Vol.21 No.6, (2010), pp.15-20 (in Japanese)

Urabe, H.; Sugawara, Y.; Ataka, M. & Rupprecht, A. (1998). Low-Frequency Raman Spectra of Lysozyme Crystals and Oriented DNA Films: Dynamics of Crystal Water. *Biophysical Journal*, Vol.74, No.3, (March 1998), pp. 1533–1540

Walther, M.; Fischer, B. M. & Jepsen, P. U. (2003). Noncovalent Intermolecular Forces in Polycrystalline and Amorphous Saccharides in the Far Infrared. *Chemical Physics*, Vol.288, No.2, (March 2003), pp. 261-268

Wieting, T. J. & Verble, J. L. (1972). Interlayer Bonding and the Lattice Vibrations of β-GaSe. *Physics Review B*, Vol.5, No.4, (February 1972), pp. 1473-1479

Yamaguchi, M.; Miyamaru, F.; Yamamoto, K.; Tani, M. & Hangyo, M. (2005). Terahertz Absorption Spectra of L-, D-, and DL-alanine and their Application to Determination of Enantiometric Composition. *Applied Physics Letters*, Vol.86, No.5, (January 2005), pp. 053903-1-053903-3.

Yamamoto, K.; Tominaga, K.; Sasakawa, H.; Tamura, A.; Murakami, H.; Ohtake, H. & Sarukura, N. (2005). Terahertz Time-Domain Spectroscopy of Amino Acids and Polypeptides. *Biophysical Journal,* Vol.89, No.3, (September 2005), pp. L22-L24

Permissions

The contributors of this book come from diverse backgrounds, making this book a truly international effort. This book will bring forth new frontiers with its revolutionizing research information and detailed analysis of the nascent developments around the world.

We would like to thank Prof. Dominique de Caro, for lending his expertise to make the book truly unique. He has played a crucial role in the development of this book. Without his invaluable contribution this book wouldn't have been possible. He has made vital efforts to compile up to date information on the varied aspects of this subject to make this book a valuable addition to the collection of many professionals and students.

This book was conceptualized with the vision of imparting up-to-date information and advanced data in this field. To ensure the same, a matchless editorial board was set up. Every individual on the board went through rigorous rounds of assessment to prove their worth. After which they invested a large part of their time researching and compiling the most relevant data for our readers. Conferences and sessions were held from time to time between the editorial board and the contributing authors to present the data in the most comprehensible form. The editorial team has worked tirelessly to provide valuable and valid information to help people across the globe.

Every chapter published in this book has been scrutinized by our experts. Their significance has been extensively debated. The topics covered herein carry significant findings which will fuel the growth of the discipline. They may even be implemented as practical applications or may be referred to as a beginning point for another development. Chapters in this book were first published by InTech; hereby published with permission under the Creative Commons Attribution License or equivalent.

The editorial board has been involved in producing this book since its inception. They have spent rigorous hours researching and exploring the diverse topics which have resulted in the successful publishing of this book. They have passed on their knowledge of decades through this book. To expedite this challenging task, the publisher supported the team at every step. A small team of assistant editors was also appointed to further simplify the editing procedure and attain best results for the readers.

Our editorial team has been hand-picked from every corner of the world. Their multi-ethnicity adds dynamic inputs to the discussions which result in innovative outcomes. These outcomes are then further discussed with the researchers and contributors who give their valuable feedback and opinion regarding the same. The feedback is then collaborated with the researches and they are edited in a comprehensive manner to aid the understanding of the subject.

Apart from the editorial board, the designing team has also invested a significant amount of their time in understanding the subject and creating the most relevant covers. They scrutinized every image to scout for the most suitable representation of the subject and create an appropriate cover for the book.

The publishing team has been involved in this book since its early stages. They were actively engaged in every process, be it collecting the data, connecting with the contributors or procuring relevant information. The team has been an ardent support to the editorial, designing and production team. Their endless efforts to recruit the best for this project, has resulted in the accomplishment of this book. They are a veteran in the field of academics and their pool of knowledge is as vast as their experience in printing. Their expertise and guidance has proved useful at every step. Their uncompromising quality standards have made this book an exceptional effort. Their encouragement from time to time has been an inspiration for everyone.

The publisher and the editorial board hope that this book will prove to be a valuable piece of knowledge for researchers, students, practitioners and scholars across the globe.

List of Contributors

Mario Alberto Gómez and Kee eun Lee
McGill University, Department of Materials Engineering, Canada

Takayuki Ebata, Ryoji Kusaka and Yoshiya Inokuchi
Hiroshima University, Japan

Paulo de Tarso Cavalcante Freire, José Alves Lima Júnior, Bruno Tavares de Oliveira Abagaro, Gardênia de Sousa Pinheiro, José de Arimatéa Freitas e Silva, Josué Mendes Filho and Francisco Erivan de Abreu Melo
Departamento de Física, Universidade Federal do Ceará, Brazil

Luciano H. Chagas, Márcia C. De Souza, Weberton R. Do Carmo and Renata Diniz
Departamento de Química, Universidade Federal de Juiz de Fora, Brazil

Heitor A. De Abreu
Departamento de Química, Universidade Federal de Minas Gerais, Brazil

Dominique de Caro, Kane Jacob, Matthieu Souque and Lydie Valade
CNRS, LCC (Laboratoire de Chimie de Coordination), 205, route de Narbonne and Université de Toulouse, UPS, INPT, LCC, Toulouse, France

Takayuki Miyamae
Nanosystem Research Institute, National Institute of Advanced Industrial Science and Technology (AIST), Japan

Masahiko Tani, Masakazu Hibi, Kohji Yamamoto, Elmer S. Estacio and Christopher T. Que
Research Center for Development of Far-Infrared Region, University of Fukui, Japan

Mariko Yamaguchi
Graduate School of Materials Science, Nara Institute of Science and Technology, Japan

Masanori Hangyo
Institute of Laser Engineering, Osaka University, Japan

Printed in the USA
CPSIA information can be obtained
at www.ICGtesting.com
JSHW011348221024
72173JS00003B/235